NEL7I1649433

D0224304

# 3D Studio Max:
# Tutorials from
# the Masters

# 3D Studio Max: Tutorials from the Masters

**Michele Bousquet**

*Wiith tutorials by:*

**Kyle McKisic**

**Alan Iglesias**

**Ted Boardman**

**Sanford Kennedy**

**Frank Delise**

**Autodesk**®

**Press**

I(T)P® An International Thomson Publishing Company

Albany • Bonn • Boston • Cincinnati • Detroit • London • Madrid
Melbourne • Mexico City • New York • Pacific Grove • Paris • San Francisco
Singapore • Tokyo • Toronto • Washington

## NOTICE TO THE READER

Publisher does not warrant or guarantee any of the products described herein or perform any independent analysis in connection with any of the product information contained herein. Publisher does not assume and expressly disclaims, any obligation to obtain and include information other than that provided to it by the manufacturer. The reader is expressly warned to consider and adopt all safety precautions that might be indicated by the activities described herein and to avoid all potential hazards. By following the instructions contained herein, the reader willingly assumes all risks in connection with such instructions. The publisher makes no representations or warranties of any kind, including but not limited to, the warranties of fitness for particular purpose or merchantability, nor are any such representations implied with respect to the material stet forth herein, and the publisher takes no responsibility with respect to such material. The publisher shall not be liable for any special, consequential, or exemplary damages resulting, in whole or in part, from the reader's use of, or reliance upon, this material.

**Trademarks**

3D Studio MAX™ and the 3D Studio MAX™ logo are registered trademarks of Autodesk, Inc.

Windows is a trademark of the Microsoft Corporation.

All other product names are acknowledged as trademarks of their respective owners.

COPYRIGHT © 1997

By Delmar Publishers Inc.
Autodesk Press imprint
An International Thomson Publishing Company
The ITP logo is a trademark under license

Printed in the United States of America

For more information, contact:

Delmar Publishers Inc.
3 Columbia Circle, Box 15015
Albany, New York 12212-5015

International Thomson Editores
Campos Eliseos 385, Piso 7
Col Polanco
11560 Mexico D F Mexico

International Thomson Publishing Europe
Berkshire House
168-173 High Holborn
London, WC1V7AA
England

International Thomson Publishing GmbH
Konigswinterer Str. 418
53227 Bonn
Germany

Thomas Nelson Australia
102 Dodds Street
South Melbourne 3205
Victoria, Australia

International Thomson Publishing Asia
221 Henderson Road
#05-10 Henderson Bldg.
Singapore 0315

Nelson Canada
1120 Birchmont Road
Scarborough, Ontario
M1K 5G4, Canada

International Thomson Publishing Japan
Kyowa Building, 3F
2-2-1 Hirakawa-cho
Chiyoda-ku, Tokyo 102
Japan

All rights reserved. No part of this work covered by the copyright hereon may be reproduced or used in any form or by any meansXgraphic, electronic, or mechanical, including photocopying, recording, taping, or information storage and retrieval systemsXwithout the written permission of the publisher.

2  3  4  5  6  7  8  9  10  XXX  02  01  00  99  98  97

**Library of Congress Cataloging-in-Publication Data**
*To Come*

# acknowledgments

A number of people pitched in to make this book and CDROM the unique product that it is.

First and foremost, I'd like to thank technical editors Larry Minton, Ken Lee, Garry Hargreaves and Paul Bloemink for their valuable comments and insights.

Thanks to Cerious Software, Inc. for allowing us to include their image browser, ThumbsPlus, on the CDROM.

Special thanks to Dan O'Leary and John Woznack at n-Space for their contribution of the Spline Length Plug-in used in Tutorial 20.

I'd also like to thank Monique Gillotti and Fran Gardino for their help in proofreading and putting together the CDROM.

This book probably wouldn't be in your hands right now if it weren't for the Herculean efforts of Kenny Citron and John Citron of KR Graphics. A galaxy-sized thanks to Kenny and John for the late nights, bad coffee, and terrible jokes that made it seem like fun and not work.

# table of contents

## Chapter 1
## Basics                                                    1

# Chapter 2
# Boolean Techniques

**47**

# Chapter 3
# Kangaroo                                               79

## How Fit Deformation Works ...................................... 81

# Chapter 4
# High Heeled Shoe                                        151

# Chapter 6
# Lighting Effects                                        **227**

# Chapter 7
# Vehicle Over Terrain

**261**

# Tutorial 18

# Tutorial 19

Tutorial 20

## Wheel and Suspension Motion

# preface

When 3D Studio MAX was released in mid-1996, a flurry of books were immediately published to meet the needs of a growing user base. Most of these books catered to beginners or specialized markets. Missing from the bunch was a book on advanced techniques in MAX.

With this in mind, I approached Autodesk Press with the idea of an advanced tutorial book. The only problem was time. As is the case with many experienced MAX users, neither I nor anyone I knew had enough time to sit down and write an entire volume. It seemed natural, then, to write the book together, with each person contributing one or more tutorials.

We were lucky indeed to work with several of the best MAX users in the world. Ted Boardman has been teaching MAX since it came out and still carries a full-time load of freelance work. Alan Iglesias barely had time to talk on the phone yet somehow managed to send me some interesting boolean techniques he had been working with recently. Sanford Kennedy was swamped with writing assignments but was making a kangaroo for fun on the side, a task that sounded perfect for an advanced tutorial. And Kyle McKisic, besides writing MAX how-tos for several magazines, has a full-time job at DreamWorks SKG Interactive. I don't think he got much sleep during the weeks he developed his tutorials, but the results were what you'd expect from Kyle — comprehensive and astonishing.

Once the book was underway, we decided to include a few beginner's tutorials to round out the information. These, along with a tutorial from Frank Delise of Kinetix, made a very complete MAX book that anyone can use.

If you're looking to become more than a beginner, this is the book for you. Power users model with booleans and fit deformation, both of which are covered in detail in this book. Scenes come alive with compound materials and intricate lighting setups. Animation can be so much more than moving objects around on keyframes — animation controllers, a new MAX feature, are unraveled here.

For those of you that read the magazine how-tos but lament the lack of detail, we've included every step in the beginner's tutorials. If you can already select and move objects, edit splines and assign materials, you can move right on to the advanced tutorials. There you'll find a host of tips on the trickier commands along with insights on their inner workings.

Several of the tutorials use ready-made models and materials. The files can be found on the CDROM included with this book. On the CDROM you'll also find models and AVIs from other users who have mastered one or another area of MAX.

We hope you enjoy this book, and can use what you learn here to create masterpieces of your own.

Michele Bousquet

# contributors

## Kyle McKisic

First introduced to 3D Studio while a senior in high school, Kyle McKisic began working in multimedia the summer immediately following his graduation in 1992. He has contributed to many interactive products including *Terminator: Rampage*, *TIME-LIFE Astrology*, and *ZORK: Nemisis* during and after his study at Ohio State University.

Kyle is a contributing writer for magazines including *Planet Studio* and *CADENCE*, and has had his work published in *Computer Graphics World*. He is a co-author of *3D Studio Hollywood and Gaming Effects* and is a certified instructor in underwater fire prevention.

Kyle's work can be seen in Autodesk's Multimedia forum on Compuserve (GO AMMEDIA), the Kinetix forum (GO KINETIX), the 3D Studio 1994 Siggraph demo reel, and often times, in the Recycle Bin on his desktop. Kyle is currently a modeler at DreamWorks SKG Interactive.

*Kyle contributed Tutorials 15 through 20.*

## Alan Iglesias

Alan Iglesias is a freelance 3D artist, animator, and production consultant living in Escondido, California. Active in the 3D entertainment industry for 5 years, and 10 years as a freelance artist, Alan has produced state-of-the-art 3D animation and imagery for clients nationwide, providing the latest looks for video games, product prototypes, broadcast commercials, industrial/training videos, and architectural previews.

Alan is a popular speaker at 3D graphics oriented conferences, coordinates the San Diego 3D Studio Users Group, and consults with Autodesk and the Yost group regarding the development of their multimedia products. He continues to seek challenging projects, primarily in the entertainment industry, hoping to utilize the latest technology to deliver the very best in 3D art, animation, and imagery.

*Alan contributed Tutorial 5.*

## Ted Boardman

Ted Boardman has operated an architectural design service since 1978 and has used AutoCAD since 1983. The majority of his work since 1991 has centered on 3D presentations and animations created with AutoCAD, 3DStudio/Max and other PC-based software.

Ted is a Kinetix Multimedia Training Specialist, and teaches 3DStudio/Max for several dealers, corporations and Autodesk Training Centers around the United States. He regularly presents seminars at national trade shows and is currently President of the Boston Area 3D Studio User Group. Ted has lived, worked and traveled extensively in Europe and Asia and has done over 28,000 miles of ocean yacht deliveries.

*Ted contributed Tutorial 6.*

## Sanford Kennedy

Sanford Kennedy is an author, animator and graphic designer based in Los Angeles. He spent many years in electrical and mechanical engineering, designing and building prototype equipment. Sanford's career in the aerospace industry began in 1965 with the Gemini and Saturn Rocket projects. He became involved in the motion picture industry when he was hired to work on special effects for Close Encounters of the Third Kind.

Since then Sanford has been involved in the special effects for 48 motion pictures and numerous commercials. Since 1992 he has been animating with 3D Studio, and most recently worked at Sony Pictures doing previsualization animation on the film *Starship Troopers* using 3D Studio MAX. He owns Sanford Kennedy Design in Los Angeles and spends his days and nights creating insects, animals, and space battles surrounded by four IBM PCs.

*Sanford contributed Tutorials 7 and 8.*

## Frank Delise

Frank Delise is currently an applications engineer for Kinetix. Frank has been using 3D Studio since 1991 and was the first instructor for 3D Studio at the State University of Farmingdale in New York. Over the last few years, Frank has created animation for the Discovery Channel, MTV and the Scientific American television series as well as several television commercials. He has also contributed to the CDROM game *Wetlands* and has created the cover art for many books on computer animation. Frank is currently involved in the development of several plug-ins for 3D Studio MAX.

*Frank contributed Tutorial 14.*

## Michele Bousquet

Michele is a freelance teacher, animator and writer. In 1990 Michele saw the wave of PC-based graphics coming, and she wasted no time in quitting her (dull) job as a database programmer to pursue an (exciting) career in 3D graphics. Three months later she won her first CADDIE award, and in 1992 she went on to work at ABC-TV in Sydney, Australia as their first in-house 3D artist.

It was there that Michele co-authored the first published books devoted entirely to 3D Studio, the 3D Studio Tips & Tricks Series. In addition to teaching all over the world, she has produced a series of videotapes on 3D artists from Europe, Australia and Russia.

*Michele contributed Tutorials 1 through 4 and 9 through 13.*

# chapter

## Basics

In this book, most buttons and panels are displayed within the instructions. Even so, the later tutorials will be much easier for you if you can find commonly used buttons easily and have a basic understanding of how MAX works.

This chapter is intended not only for those who have never worked with MAX, but for users who are struggling with some of its basic concepts.

If you've never used MAX before, it is imperative that you do the practice exercises in this chapter before going further. Please don't make the mistake of jumping in and trying to figure it out as you go along. This scenario always ends the same way, with protests that MAX is "too hard to use".

Learning the fundamentals is a must before you can work with MAX's more complex features. I have heard from many power users that MAX is an intuitive program, that anyone can learn how to use it just by playing around with the buttons. Such people have simply forgotten their early days with MAX.

You really do have to do these exercises and work with the commands until you understand the flow of the program. Then the next time someone asks you if MAX is hard to learn, you can say with confidence that it's a piece of cake — once you know the basics.

If you've used MAX but you're having trouble selecting objects and getting to the commands you want, you can skip the first exercise. However, I recommend all these exercises to anyone who doesn't quite get how MAX works.

# Practice Exercise 1
# Navigating MAX

At first glance, MAX's screen can look pretty daunting. This tutorial's goal is to familiarize you with the screen layout and to make it easier for you to find commands and buttons later on.

It is assumed that you have MAX up and running on your computer.

## Step 1. Screen Layout

Look over the screen layout. The work area takes up most of the screen. Four viewports give you a top, front, side and perspective view of your work. Only one viewport is current at any time. The current viewport is indicated by a white bounding box. Click in each viewport to see how viewport selection works.

The toolbar along the top of the screen holds several buttons. Move your cursor over a button and leave it there without clicking. After a moment, a small label appears telling you the name of the button.

When you click on the large gray buttons at the upper right, more buttons and commands appear. The buttons along the bottom of the screen control snap and display.

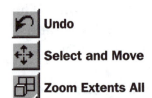
*If your screen resolution is 800x600, you won't be able to see all the buttons at the top of the screen. However, you can scroll the button display to the left or right at any time to get access to all the buttons. To do this, move your cursor over the toolbar until it changes to a small hand. Click and drag to move the toolbar and reveal more buttons.*

### Step 2. Toolbar

Use Figure 1-1 to help you locate the following buttons.

**Undo**

**Select and Move**

**Zoom Extents All**

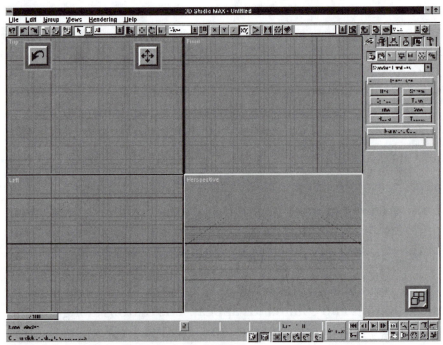

*Figure 1-1. Locations of commonly used buttons*

These three buttons are your friends. You'll be spending a lot of time with them, so you should get to know them well.

### Step 3. Command panels

At the right of the screen are six buttons called command panels. Move the mouse over each one to see its name. In particular, find the Create button and the Modify button. You'll use these commands the most when working with MAX. As you click on each panel, the area under the button changes to show the buttons and entry fields available for that command.

Every command panel has one or more rollouts. Each rollout has a title with a plus or minus sign to the left of it. If a minus sign appears, it means the rollout is expanded. A plus sign means the rollout is collapsed. You can click on the rollout name bar to expand or collapse it. Rollouts make it possible to have a large number of buttons and entry fields under any one panel.

By default, the Create panel is already chosen. The Create panel is displayed. Below the panel buttons, eight more buttons appear. These are subdivisions of the Create command. The first button, Geometry, is already pressed. Move your mouse over each of the eight buttons to see what each one is for. The first two buttons, Geometry and Shapes, are the two you'll use the most when creating models.

Under the Object Type rollout, find the button labeled Teapot and click the button. Several rollouts appear. If you want to get to the last rollout, there are two ways to go about it. One is to move the cursor over the panel until the small hand appears, then click and drag to move the panel upward. The other way is to collapse all rollouts except the one you want to see.

### Step 4. Making an object

Click on the Box button. Click and drag in any viewport to create a rectangle. Then let go and move your mouse to create the height of the box. Click to set the height.

Click on Zoom Extents All.

When you clicked on Box, a rollout called Parameters appeared. As you created the box, the Length, Width and Height entries changed accordingly.

After you create an object, you can change its parameters. Type in new Length, Width and Height values for the box and note the change on the screen.

Also adjust the segment values for the box, and note the change on the screen. The segment values are labeled Length Segs, Width Segs and Height Segs.

*If you change a value by typing in a new value, the new value won't take effect until you press <Enter> or click on another entry on the panel. Spinners take effect right away.*

You can change these values by typing in a new value, or by clicking and dragging on the small arrows to the right of each entry field. These small arrows are called spinners.

Click on Sphere, Teapot or any of the other buttons under Object Types. Make more objects until your screen is full of them. Practice changing an object's parameters just after you make it.

Click on Zoom Extents All to see all the objects in all viewports.

When you're making an object, you can cancel the operation at any time by clicking your right mouse button.

When you create an object, its tripod axes appear at the center of the object. These XYZ axes are used to orient the object in 3D space. The axes might seem to clutter up your screen, but you'll soon get used to them. Axes also come in very handy when modifying objects, as you'll see later on.

As you create each object, it is automatically named using the object type. For example, the first box you create is named Box01. The name appears in the rollout at right. You can change the object name at any time, either while creating the object or later on when modifying it. To change the object name, highlight the name, enter a new name and press <Enter>.

Note that just above the Object Type rollout is a pulldown list that currently reads Standard Primitives. You have just made several primitives. The term primitive is used in many 3D packages to indicate a basic type of object.

To see the other types of objects, click on the down arrow next to Standard Primitives. Don't concern yourself just yet with what they are. You'll be working with these other types of objects over the course of this book, and you'll learn about them as you go along.

Save your work as BASIC01.MAX. To do this, pull down the File menu from the top left of the screen. Choose Save. Enter the name BASIC01. The extension MAX will automatically be added to the filename.

## Practice Exercise 2
## Selecting and Transforming Objects

In this tutorial you'll learn the most commonly used ways to select and modify objects.

If you've done the previous tutorial, load the file BASIC01.MAX if it's not already on your screen. To do this, Choose Load from the File menu and pick BASIC01.MAX from the list that appears.

If you haven't done the previous tutorial, go to the Create panel and make a few primitive objects such as boxes and cylinders. These objects will be sufficient for this tutorial.

### Step 1. Basic Selection

Before you can do anything with an object, you have to select it.

On the toolbar, locate the Select button. Click on the button to turn it on.

When you move your cursor around the screen, you will see it sometimes turn into a small ✚ symbol. This indicates that the cursor is over a selectable object.

*Whenever you click in a different viewport, the selected object is no longer selected. Use right-click to switch viewports without losing your selection.*

Move the cursor over your objects until you see a ✚ cursor. Click to select the object. The selected object turns white. This is the simplest way to select an object.

You can select multiple objects in a few different ways. One way is to use the <Ctrl> key. Select an object, then hold down the <Ctrl> key and click on another object. Both objects are now selected. You can select as many objects as you like in this way.

A second way is to draw a bounding box around the objects. Go to an area of the viewport where no objects lie. The cursor changes to an arrow. Click and drag the cursor to make a bounding box around several objects, then let go. The objects that sit inside the bounding box are selected.

## Step 2. Transforms

Once an object is selected, you can modify it. One way to modify an object is to use a transform. MAX has three transform functions: move, rotate and scale.

MAX's transforms can be accessed via buttons on the toolbar.

 **Select and Move**

 **Select and Rotate**

 **Select and Uniform Scale**

You can use these functions to select and transform an object in one step. To use a transform, click on the transform button to turn it on, and click and drag on an object. The object doesn't have to be selected before you begin. As soon as you click on an object, it becomes the new selected object.

Practice using these buttons on the objects you created. You can work in any of the four viewports. Note that sometimes you'll have to move the mouse from side to side to get the result you want, while other times you'll need to move it up and down.

The Select button you used in Step 1 isn't useful for anything more than selecting objects. For this reason, most users keep Select and Move turned on while they're working rather than Select. In this way, you can either select and move, or just select, without having to click on another button in between.

*You can use a transform when you simply want to select an object and not transform it. For example, when Select and Move is turned on, you can just click on an object to select it without moving it.*

### Step 3. Flyouts

Some buttons are really two or more buttons combined. Click and hold the Select and Uniform Scale button. Three buttons appear underneath it, each a different type of scale. Move the cursor over the second button and let go. The toolbar now shows the new button for the type of scale you selected.

This type of pulldown is called a *flyout*. Many buttons in MAX have flyouts. A button with an available flyout has a small triangle at the lower right of the button.

There are four flyouts you'll use most often. Locate each of the flyouts on the toolbar. Click and hold each flyout and look at the names of the buttons on each one. Don't be concerned with their functions, just make sure you can find them when you need them.

Note that a flyout displays the currently selected button as well as the other selections.

**Select and Uniform Scale**

**Select and Non-uniform Scale**

**Select and Squash**

**Zoom Extents All**

**Zoom Extents All Selected**

**Rectangular Selection Region**

**Circular Selection Region**

**Fence Selection Region**

Restrict to XY Plane

Restrict to YZ Plane

Restrict to ZX Plane

## Step 4. Restrict Axis buttons

When you're working with a transform, you can restrict the change to one axis with the buttons shown below.

 Restrict to X

 Restrict to Y

 Restrict to Z

To try out the axis constraints, click and hold Select and Uniform Scale ⬛ to see the flyout. Choose Select and Non-uniform Scale. ⬛ Click in the Top viewport. Click on the Restrict to X button. ⬛ In the Top viewport, scale an object. The object scales only along the X axis.

Practice working with the axis constraints and the scale and rotate transforms. See if you can work out the relationship between the axis constraints and the object's tripod axes.

> *When modifying an object with transforms, do your work in this order:*
> *1. Pick the viewport.*
> *2. Pick the transform.*
> *3. Pick the restriction for the axis.*
>
> *Choosing a new transform changes the axis restriction to the last one used for that transform. Doing your work in the order listed above will help you avoid extra mouse clicks.*

### Step 5. Zooming

The buttons at the lower right of the screen are the zoom controls. Try out each of these buttons.

The buttons you'll use most often are the Zoom Extents buttons. These buttons change one viewport or all viewports to show all objects.

 **Zoom Extents**

 **Zoom Extents All**

These two buttons also contain flyouts for performing a Zoom Extents on selected objects.

 **Zoom Extents Selected**

 **Zoom Extents All Selected**

With the Zoom and Zoom All buttons, you click and drag in a viewport to change the view. Zoom All changes all the viewports at the same time.

 **Zoom**

 **Zoom All**

With Region Zoom, you click and drag to create a bounding box. The view zooms to the bounded view.

 **Region Zoom**

The Pan button lets you click and drag to shift the view in the direction of the drag.

 **Pan**

Arc Rotate rotates the view in any direction. When you choose Arc Rotate, a circle appears on the screen. As you move the cursor to different parts of the circle, the cursor changes to various types of curved arrows indicating the direction of rotation. Move the cursor to display a curved arrow, then click and drag to rotate the view.

 **Arc Rotate**

When you rotate the view, the viewport label changes to User. This indicates that the view has been customized.

The Arc Rotate button has a flyout which contains the Arc Rotate Selected button. This button causes the view to use selected objects as the center of rotation.

 **Arc Rotate Selected**

You can change the display so that one viewport fills the entire screen. To do this, activate a viewport and click Min/Max Toggle.

Click the same button again to return to the four-viewport display.

### Step 6. Advanced selection

Right now, when you select objects by drawing a bounding box around them, all objects touched by or inside the bounding box are selected. You can change the selection mode so that instead, only the objects that are completely inside the box are selected.

The current selection mode is called *crossing* selection. The other selection mode is called *window* selection.

The Window/Crossing Selection toggle button can be used to change the mode of selection. Locate the button at the lower center of the screen.

 **Crossing Selection**

 **Window Selection**

Practice using both types of selection on your objects.

If you like, save your work as BASIC02.MAX.

## Practice Exercise 3
## Modifying Objects

Once an object is created, it can be modified (changed) in many ways.

### Step 1. Modify panel

Load the File BASIC01.MAX. Select any object. Click on the

Modify panel button at the upper right of the screen.  The Modify panel appears.

There are many different types of Modify panels in MAX. The Modify panel that appears depends on the type of object selected. For a box, the Modify panel shows creation parameters such as length, height and width. For a cylinder, you'll see the radius and height.

Change some of the entries and watch the screen to see how the changes affect the object.

Select another object. The Modify panel changes to correspond to the newly selected object. Change one or more entries on the panel.

Continue selecting and changing objects until you have a feel for how the Modify panel works.

### Step 2. Modifiers

Under the Modify panel there are several buttons called modifiers. These are functions that change an object's shape.

To try out a modifier, select an object. Click on Bend. The Modify panel changes to show the Bend parameters. Increase the Amount spinner and watch the screen to see the effect. Change the Bend Axis and see how that affects the bend. See if you can work out the relationship between the Bend axis and the tripod axis of the object.

Try the Taper and Twist modifiers on different objects. You can also click the More button to see a list of more modifiers. Try the Skew modifier. Note that with these modifiers, nothing happens to the object until you change the Amount spinner.

### Step 3. Modifier stack

When you apply a modifier to an object, it goes onto the Modifier Stack. This stack is a list of all the modifiers applied to an object.

After you're finished applying a Bend modifier to an object, for example, you can click on Taper and modify the object further. Then both the Bend and Taper modifiers are on the stack.

*Only modifiers picked from the Modify panel appear on the stack. Parameter changes entered in the Modify panel, and other modifications such as rotating the object with Select and Rotate, do not appear on the stack.*

To see an object's stack, click on the Edit Stack button under the Modifier Stack rollout. The lowest entry on the stack is the creation of the object. Modifiers appear on the stack in the order in which they were applied, from bottom to top.

If you later decide that you don't like the effect a modifier has on your object, you can delete it from the stack at any time.

### Step 4. Gizmo

Many modifiers give you access to a Sub-Object level for further control of the modification. To see how this works, apply a Bend modifier to an object. Change the Amount spinner to bend the object. Now click on the Sub-Object button. The name Gizmo appears next to the Sub-Object label.

A *gizmo* controls the effect of the modifier on the object. Look at the selected object. A box appears around the object indicating the location of the gizmo. The gizmo also has large crosshairs indicating the center of the gizmo.

Click Select and Move and move the gizmo in each viewport so the center crosshairs align with the bottom of the object. Note the change in the bend as you move the gizmo.

When you're done modifying the gizmo, click on Sub-Object to turn it off.

### Step 5. Sub-Object level

Other modifiers have different Sub-Object levels. Select an object and choose the Edit Mesh modifier. The Sub-Object level is automatically turned on, and Vertex is the current level. The object display changes to show all the vertices on the object.

Click Select and draw a bounding box around some of the vertices. The vertices turn red to indicate they are selected. Click

Select and Move. Move the vertices.

Try using different transforms on the vertices. When you've finished, turn off Sub-Object.

If you like, save your work as BASIC03.MAX.

# Practice Exercise 4
# Shapes

*Under the Cre-ate panel, the Shapes button displays a panel with a selection of 2D shapes, while the Geometry button leads you to 3D object types.*

Shapes are 2D objects that can be used to make 3D objects. Examples of shapes are circles and rectangles. Do not confuse a circle with a cylinder, or a rectangle with a box. Circles, rectangles and other 2D shapes are flat with no dimensions, and will not render. However, they are very useful for creating 3D objects that will render.

## Step 1. Create shapes

Under the Create panel, click on the Shapes button. Click on Circle. Draw click and drag to draw a circle in any viewport.

Choose some of the other shape buttons such as Rectangle and Ellipse to make more shapes.

Click on Line. Click and drag to start the line, then click points to create the line. Right-click when the line is done.

Note that as you create each shape, a name is automatically assigned. The name refers to the type of shape created, such as Circle01 and Line02. As with objects, you can change the name of a shape at any time. To change the shape name, highlight the old name, enter a new name and press <Enter>.

When you have filled the screen with shapes, go on to the next step.

## Step 2. Edit vertices

Clear the screen by choosing New from the File menu. Click OK to clear the screen. Create an ellipse in the Top viewport.

Click on Zoom Extents All.

Shapes are defined by vertices. A vertex is a defining point on the shape. A vertex is denoted by a small + on the shape.

You will often be required to edit shapes for your modeling needs. The easiest way to do this is to modify individual vertices on a shape.

Select the ellipse. Under the Modify panel, click on the Edit Spline button. A new rollout appears.

Shapes can also be modified at the sub-object level. The Vertex sub-object level is on by default as soon as you click on Edit Spline.

A vertex can be a corner vertex or a smooth vertex. A corner vertex makes a sharp edge, such as the corners on a rectangle. A smooth vertex makes a rounded edge like the side of a circle.

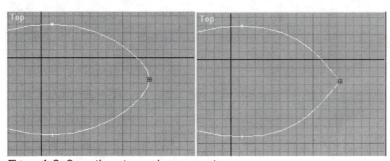

Figure 1-2. Smooth vertex and corner vertex

Both types of vertices can have handles. These handles can be moved individually to change the shape.

**Because Restrict to XY Plane is turned on, the vertices are scaled only in the Top viewport's XY plane.**

Click on Select and Move, 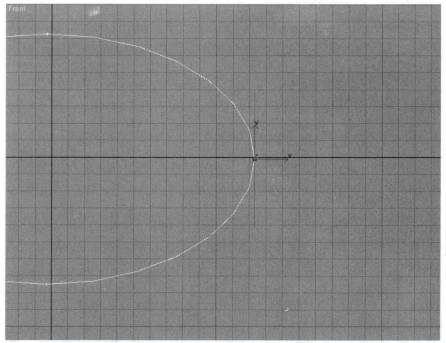 and turn on Restrict to XY Plane. In the Top viewport, click on any vertex on the ellipse. The selected vertex turns red, and two green *handles* appear around it.

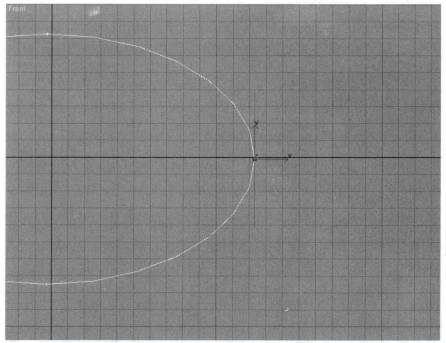

Figure 1-3. Ellipse vertex with handles

Click and drag on the vertex to move to another location. Click to set it down. The parts of the shape around the vertex retain their rounded shapes even when the vertex is moved.

Click and drag on one of the green handles and see what happens. The shape changes according to the movement of the handle.

Now right-click on the vertex. A list appears.

| |
|---|
| **Move** |
| **Rotate** |
| **Scale** |
| Select Children |
| Deselect Children |
| **Smooth** |
| **Corner** |
| √ **Bezier** |
| **Bezier Corner** |

The last four items on the list are vertex types. Change the vertex to each different type and see how it affects the shape.

Bezier and Bezier Corner are the two types of vertices that use handles to control the shape of the curve. At first, there appears to be no difference between Bezier and Bezier Corner. The difference isn't apparent until you move one of the handles. With Bezier Corner, you can move handles individually to make a corner of any angle. With Bezier, the handles always move together to keep the line between the handles tangent to the shape.

Practice changing vertex types and changing the shape until you feel comfortable with editing vertices.

## Step 3. Edit segment

The sub-object levels for shapes are Vertex, Segment and Spline. A segment is the line between two vertices.

Most of the time, any editing you do to segments will be to change a segment from curved to straight, or vice versa.

To do this, click on the pulldown list next to Sub-Object and choose Segment. Click on any segment on the ellipse. The segment turns red. Right-click on the selected segment. A list appears.

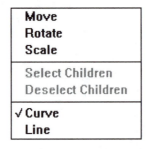

**Move**
**Rotate**
**Scale**

Select Children
Deselect Children

√ Curve
Line

*Once a segment is changed to Line type, you can't always get the same curve back by changing back to Curve type.*

Click on Line to change the segment to a straight line. To change the segment back to a curve, right-click on the segment again and choose Curve.

### Step 4. Edit spline

The third type of sub-object is Spline. A spline can be an open or closed collection of vertices and segments. Because a spline can also be a complete shape, many MAX users use the terms shape and spline interchangeably.

A frequent use of the Spline sub-object level is for putting together two halves of a shape. To see how this works, you'll create a circle and change it into a heart.

Create a circle in any viewport. Under the Modify panel, click Edit Spline. Move the vertices and adjust their handles to create a heart shape.

Figure 1-4. Heart shape

*To move one handle at a time, hold down the <Shift> key and drag the handle.*

*To move the top and bottom vertices of the circle up and down only, turn on Restrict to Y before moving the vertices.*

You'll probably find that it's difficult to get the two sides of the heart to look exactly the same. To fix this, you'll delete half of the heart shape, then copy and mirror the remaining spline.

Change the Sub-Object level to Segment. Select the two segments on one side of the heart. The segments turn red to show they are selected. Press the <Delete> key to delete the selected segments.

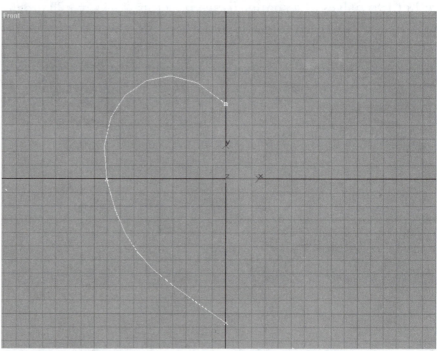

*Figure 1-5. Half of heart shape deleted*

Change the Sub-Object level to Spline. Click on the heart shape to select it. Look under the Edit Spline rollout and locate the Mirror button. Turn on the Copy checkbox, then click Mirror. A mirrored copy of the spline appears.

Move the mirrored copy so the ends of the splines touch. A dialog box appears.

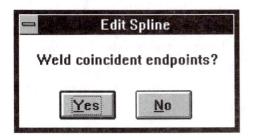

*Vertices will only be welded if they are very close together. If vertices on the heart do not weld together, move one vertex on top of the other to weld them.*

Click Yes. The two splines are welded together to make one symmetrical spline.

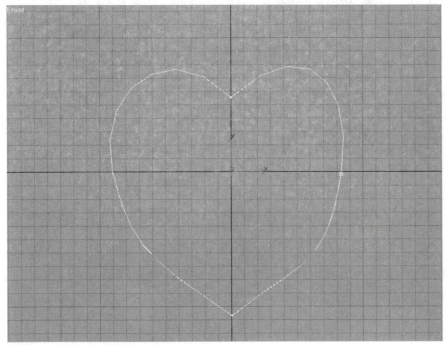

*Figure 1-6. Symmetrical heart*

### Step 5. Refine

Sometimes you will need to add more vertices to a spline to get the shape you want. You can do this with the Refine option.

Change the Sub-Object level to Vertex. Under the Edit Vertex rollout, click Refine. Move the cursor over a shape. The cursor changes to include two short lines, indicating that you are in Refine mode.

Click on any shape to place a vertex. The new vertex can be edited like any other vertex. Click again to place another vertex.

When you're done placing vertices, you must get out of Refine mode before you can do anything else. The easiest way to do this is to click Select and Move. Practice using Refine to add vertices, and click Select and Move when you're done.

### Step 6. Line

So far we've worked with only closed shapes. A line can be an open or closed shape of any kind. Use the Line type shape to create a straight line, or to create odd shapes from scratch.

> **TIP** *The Line tool is the one used most often in this book for creating shapes. Practice making lines with straight and curved vertices until you are very comfortable doing so. This will make your work with later tutorials go much more quickly.*

Under the Create panel, click Line. Click in a viewport to start the line, then move the cursor and click again. Continue clicking in different areas to create a shape.

If you want to close the shape, finish the line by clicking near the starting point of the line. If you want to leave it open, right-click to end the line.

When creating a line, you can make the vertices curve automatically. When you click to create a vertex, hold the mouse button down and move it around. The line becomes curved around the vertex. Let go of the mouse button and move the cursor to create the next vertex.

## Step 7. Extrude

Once a shape is complete, you can use it to make a 3D object. The quickest way to do this is with the Extrude modifier.

Select a shape. Go the Modify panel and click Extrude. Change the Amount spinner and watch the shape turn into a 3D object. This 3D object will show up in renderings.

If you like, save your work as BASIC05.MAX.

## Practice Exercise 5
## Lofting

Lofting is MAX's most useful modeling technique. With lofting, a 2D shape is passed along a 2D path to define a 3D object.

### Step 1. Making a path

For the lofting path, a straight line is the most commonly used shape. Create a straight line for the path with the Line command under the Create panel. Right-click on each vertex to make sure they're both Corner type.

### Step 2. Shape

Create a closed shape of any kind, such as a circle or rectangle.

### Step 3. Loft

Select the path. Under the Create panel, click Geometry. On this panel is a pulldown list that currently reads Standard Primitives. Change the pulldown selection to Loft Object.

A new panel appears. Under the Object Type rollout, click the Loft button. Click on Get Shape, then click on the closed shape you just created. Click on Zoom Extents All. A copy of the shape appears at the end of the path. The object has been lofted, but you can't see it yet.

Expand the Skin Parameters rollout. Turn on the Skin checkbox. This option makes the "skin" of the loft object appear in all viewports.

## Step 4. Modify loft object

With the loft object selected, go to the Modify panel. You can modify the loft object as you would any other object.

If you like, save your work as BASIC06.MAX.

## Tutorial 1
## Travel Mug

In this tutorial, you'll build a cup. Many of the functions used later in the book appear in this tutorial. This simple tutorial will help you become familiar with how MAX works, and will make your later work go much more quickly.

Figure 1-7. Travel mug

> **TIP** *Before starting any project, think about how you will use the tools you have to create the model. It's also helpful to have the object itself, a similar object or a photo of the object within view while you're modeling. This travel mug is a modified version of a mug I use every day, and I had it sitting on my desk as I made the model.*

Before we begin, let's take a few moments to think about how this mug should be created.

The cup could be created with a tube, but then it would have no bottom. You could place a flat cylinder at the bottom to close the cup, but there would be a visible seam where the cylinder meets the tube. To avoid a seam, you'll use a cylinder for the mug, and modify vertices to make it into a seamless cup.

The handle will be created with an ellipse lofted along a curved path. The entire mug will then be tapered.

Before starting this tutorial, reset MAX. To do this, choose Reset from the File menu. When asked if you really want to Reset, answer Yes.

## Step 1. Create cylinder

Create a cylinder in the Top viewport. Give the cylinder the following parameters.

| | |
|---|---|
| **Radius** | **50** |
| **Height** | **180** |
| **Height Segments** | **1** |
| **Cap Segments** | **3** |
| **Sides** | **18** |

Change the name of the cylinder to Mug. To do this, highlight the name Cylinder01 that appears under the Name and Color rollout. Enter the name Mug and press <Enter>.

Click on Zoom Extents All to see the cylinder in all viewports.

## Step 2. Change viewport display

Change the Perspective viewport to Smooth + Highlight display. To do this, right-click on the viewport name Perspective. A list appears.

| |
|---|
| √ Smooth + Highlight |
| Faceted + Highlight |
| Wireframe |
| √ Show Grid |
| Show Background |
| Show Safe Frame |
| Texture Correction |
| Disable View |
| Views ▶ |
| Swap Layouts |
| Undo |
| Redo |
| Configure... |

Choose Smooth + Highlight from the list. You can now see a shaded cylinder in the Perspective viewport.

### Step 3. Select vertices

Right now, the cup is a solid cylinder. To make it hollow, you'll select vertices at the top of the cylinder and pull them down.

Under the Modify panel, click Edit Mesh. The Vertex sub-object level is automatically activated.

Click and hold on the Rectangular Selection Region button

to display the flyout. Choose Circular Selection Region.

In the Top viewport, click at the center of the cylinder. Move the cursor outward until the bounding circle includes just the center ring of vertices. Let go of the mouse to set the selection.

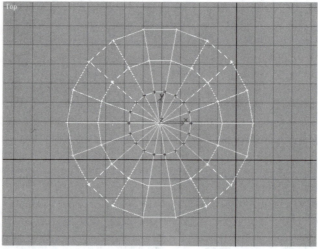

*Figure 1-8. Selected vertices*

In the Front and Left viewports, you can see that you've also selected vertices at the bottom of the cylinder. These vertices must be removed from the selection set.

Click and hold on the Circular Selection Region button and

change back to the Rectangular Selection Region.

Right-click in the Front viewport to activate it. Move the cursor to the lower left of the Front viewport. Hold down the <Alt> key. The cursor changes to include a small minus sign. This minus sign indicates that you can now remove vertices from the selection set.

Holding the <Alt> key down the entire time, draw a bounding box around the bottom of the cylinder. The vertices at the bottom of the cylinder are no longer selected.

Check to make sure the vertices at the top of the cylinder are still selected. If not, it probably means you didn't hold down the <Alt> key throughout the entire deselection process. If so, repeat this step.

### Step 4. Modify vertices

Click Select and Move, ⬛ and turn on Restrict to Y. ⬛ In the Front viewport, move the vertices down close to the bottom of the cup.

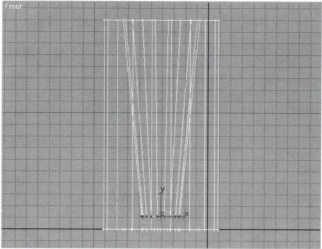

Figure 1-9. Moved vertices

To see the hole you just made in the top of the cup, right-click on the Perspective viewport. Click Arc Rotate. ⬛ Move the cursor to the bottom node on the green arc controller. Click and drag to change the display to look down at the top of the cup.

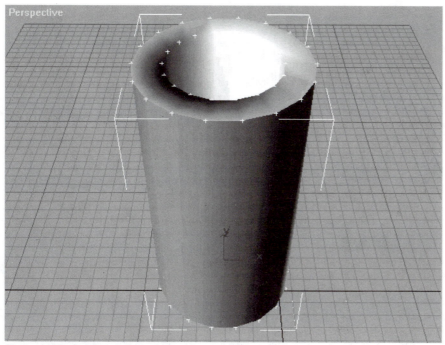

*Figure 1-10. Overhead view of cup*

Right-click in the Top viewport to activate it. Click Select and Uniform Scale.  In the Top viewport, move the cursor until a small icon similar to the button you just chose appears on the screen. Click and drag to scale the vertices to 200%. Watch the status bar at the bottom of the screen to see the percentage.

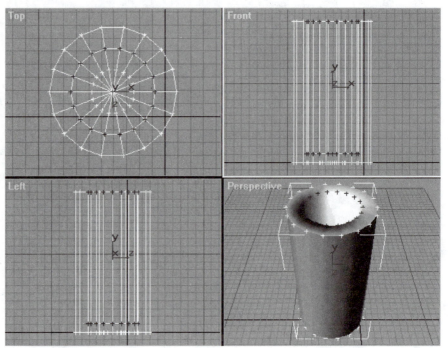

*Figure 1-11. Scaled vertices*

### Step 5. Modify cup

Now you'll modify the cup so it's not so thick from the inside to the outside.

Click on the Top viewport, and click the Select button.

Change to Circular Selection Region. Select all vertices except the outermost ring of vertices.

Right-click on the Front viewport. Change to Rectangular Selection Region. Deselect the vertices at the very bottom of the cup. Your selection set should look like the one in Figure 1-12.

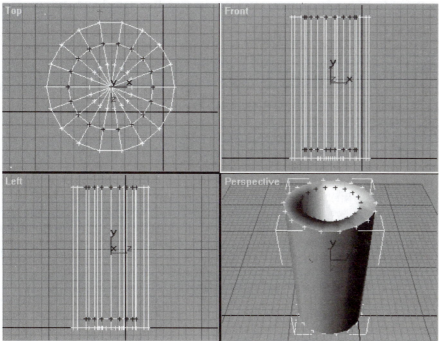

*Figure 1-12. Selected vertices*

Right-click on the Top viewport. Click and hold Select and Uniform Scale and choose Select and Non-uniform Scale from the flyout.

In the Top viewport, scale the vertices to 130%.

The cup itself is now complete. Next you'll create a handle for the cup.

### Step 6. Handle

The handle for the cup will be created with an ellipse lofted along a curved path. The path will be created with the line tool.

Under the Create panel, click Shapes. Click the Line button.

In the Front viewport, create a curved line with three vertices to represent the handle. Start the line inside the cup. Click to set the first point. When you reach the spot for the second point, click and hold to make a curved line as shown in Figure 1-13. Click to set the third point just inside the cup.

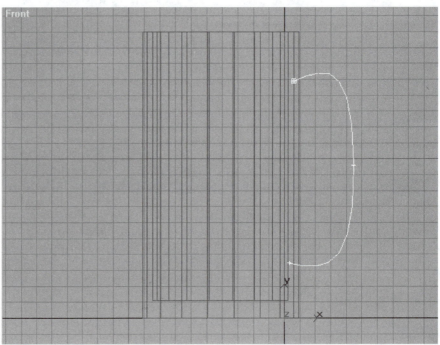

*Figure 1-13. Line for cup handle path*

Name the line HandlePath.

If the line doesn't look right on the first try, go to the Modify panel and click Edit Spline. The Vertex sub-object level is automatically turned on. Change the types of the two end vertices to Corner, and change the middle vertex type to Bezier. Pull the handles on the middle vertex so they are very long and straight, as shown in Figure 1-14.

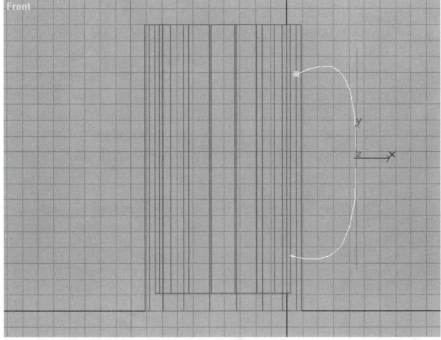

Figure 1-14. Bezier handles for middle vertex

Click Select and Move. 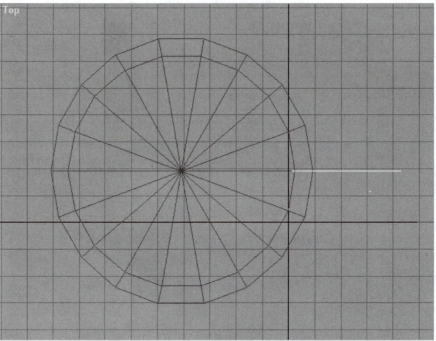 In the Top viewport, move the path so it's aligned with the center of the cup.

Figure 1-15. Path aligned with center of cup

Under the Create panel, click Circle. Create a small circle in the
Front viewport.

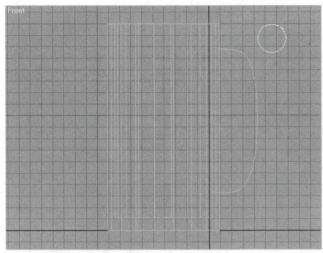

Figure 1-16. Small circle

Select HandlePath. Under the Create panel, click Geometry.
Choose Loft Object from the pulldown menu. Click the Loft but-
ton, then click Get Shape. Click on the circle. Under the Skin
Parameters rollout, turn on the Skin checkbox.

The handle appears in all four viewports. Change the name of
the handle to Handle.

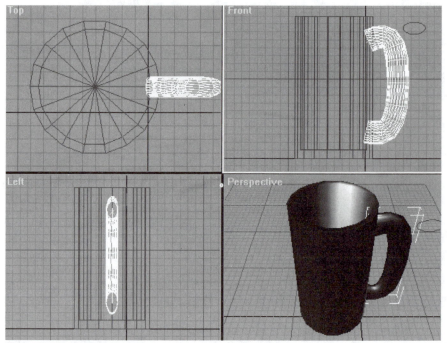

Figure 1-17. Cup and handle

The handle is not completely smooth. To smooth it out, you need more steps on the path. Under the Skin Parameters rollout, change Path Steps to 12.

### Step 7. Taper cup

Next you'll apply a Taper modifier to the cup.

Select the cup. Click Select and select the cup. Under the Modify panel, click Taper.

Change the Amount spinner and watch the taper. You want the cup to taper from the bottom, but it's tapering from the top. You'll have to change the Taper Gizmo to make the cup taper from the bottom.

Change the Amount back to zero. Click on Sub-Object. The Gizmo level is automatically selected.

Click Select and Rotate. 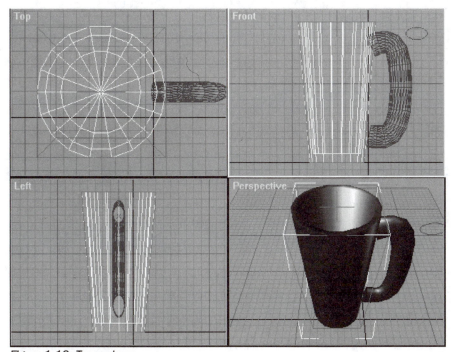 Turn on the Angle Snap Toggle. Click in the Front viewport, and make sure Restrict to Z is on. In the Front viewport, rotate the gizmo by 90 degrees. Watch the status line at the bottom of the screen to see how far the gizmo is rotated.

Click Select and Move and turn on Restrict to Y. Move the gizmo up in the Front viewport so it sits over the cup.

Turn off Sub-Object. Change the Amount to -0.3. The cup is now tapered at the bottom.

*Figure 1-18. Tapered cup*

43

### Step 8. Rotate handle

The handle must be rotated and moved to fit against the cup. Activate the Front viewport, and click on the handle to select it.

Click Select and Rotate ⟳ and make sure Restrict to Z is on.

Z Rotate the handle by -2 degrees in the Front viewport.

Click Select and Move ✛ and turn on Restrict to X. X In the Front viewport, move the cup handle to the left until it sits just inside the cup.

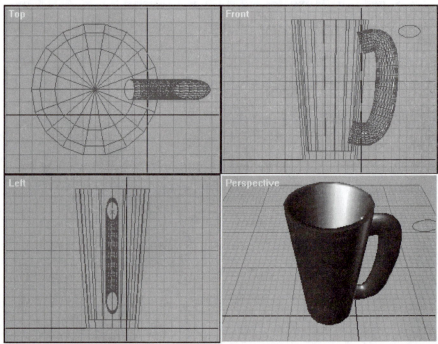

*Figure 1-19. Moved handle*

The cup is now complete. Save your work as TRAVMUG.MAX.

Feel free to experiment with this model, using different modifiers on the cup. You can also edit the vertices of the original ellipse to change the shape of the handle.

## Modes of Operation

MAX differs from 3D Studio R4 and other programs in that it is mostly modeless. This means you can jump around between operations, even if the operation isn't complete. For example, while you're in the middle of assigning materials, you can pop over to the Modify panel and change the object's parameters. The Material Editor window can stay on the screen the entire time. When you're done with the parameters you can go right back to your materials and pick up where you left off.

Until you get used to it, modeless operation can be confusing. Many operations have no finite ending; they remain "open" indefinitely. On the other hand, some operations definitely have to be closed before you can move on.

Here are some of the rules for navigating MAX's modeless system. If you're new to MAX, these rules might not make any sense right now. If you find later that you're having trouble selecting objects or getting MAX to do what you want, refer back to these rules.

### Keeping Select and Move on

Most of the time you'll work with Select and Move turned on. After you click Select and Move, it will stay turned on through various operations — editing splines, extruding, lofting.

Sometimes you'll be in the middle of an operation and will want to select another object, but it will seem impossible to do so. If this happens, check if Select and Move is still turned on. Chances are, it's not. You can usually get out of the current operation just by clicking Select and Move. This will end the operation and let you select another object.

### Select with Sub-Object on

Some commands under the Modify panel have a Sub-Object button. When you turn this button on, you can't choose another object until you either turn the button off or pick another command panel altogether.

You can't simply pick the object. You have to either turn off Sub-Object, click on another command panel such as Create or Display, or click Select and Move  before you can select another object.

### Selecting after specific commands

Modeless operation means you can perform operations in any order you choose. There's no set order for commands. Immediately after clicking on the Taper button, for example, you don't have to stay with the Taper options. You can just as easily change your mind and choose another object for modification.

This is true for almost all MAX commands. With some commands, however, you must perform a specific task immediately after clicking the button.

One example is lofting. When you click the Get Shape or Get Path button, MAX expects you to pick the shape or path right away. You can't decide you don't want to loft after all, and just click on an object expecting to modify it.

If you choose one of these commands and decide not to go through with the operation, the best way out is to click one of the selection buttons such as Select ▲ or Select and Move. ✛

# chapter 2

## Boolean Techniques

In MAX, you can use two objects to make one object, combining them, subtracting them or obtaining their intersection. These functions are performed with boolean operations.

Booleans are standard tools for professional MAX users. Most complex jobs require several boolean operations to get the right result. Many MAX users try using booleans but have no success, and subsequently give up. Booleans can be made to work, but certain rules must be followed.

In this chapter, different ways of using booleans are explored. Tutorials 2, 3 and 4 cover all the different ways a boolean can go wrong, and tell you how to correct them. Tutorial 5 shows how to use boolean intersection to improve the quality of your models.

## How Booleans Work

Booleans work with two objects. Objects used for a boolean operation are called operands. During a boolean operation, these objects are referred to as Operand A and Operand B.

There are three types of boolean operations. Subtraction subtracts one object from another. Union puts the two objects together as one object, removing all overlapping faces. Intersection results in an object containing only the intersecting faces of the two objects.

In general, booleans work with the following procedure.

1. **Create two objects.**
2. **Move one or both objects so they intersect.**
3. **Select one object. This is Operand A.**
4. **Under the Create panel, click Geometry and choose Compound Objects from the pulldown list.**
5. **Click Boolean.**
6. **Under the Parameters rollout, pick an operation type.**
7. **Click Pick Operand B, and click the other object.**
8. **A new object appears based on your boolean operation.**

A new object created with a boolean operation is called a *boolean result*. When performing a boolean operation, you can choose to instance one of the operands. Later on, you can change the operand and automatically update the boolean result.

It all sounds so simple, and much of the time, it is. But there are certain circumstances under which booleans won't work.

In order for a boolean operation to be successful, certain rules must be followed.

An operand cannot be two or more attached objects. If you have two objects you want to use as one operand, don't attach them together first. You can either perform a boolean union to put them together first, or use each separately for two boolean operations.

An operand cannot be an open object. In the first part of a boolean operation, MAX looks for the outside of each operand. An open object is missing its "outside" in some parts. An example of an open object is a cylinder with its top removed, or an extruded spline with no cap.

Operands cannot have coincident faces. When two faces occupy the same space, they are said to be coincident. Boolean operations don't work when the two operands have coincident faces. This problem is usually very difficult to detect. In many cases, the only way you can tell that there are coincident faces is that the boolean operation refuses to work. There are many solutions to this problem, some of which are explored in Tutorials 2, 3 and 4.

There are some situations that commonly result in coincident faces.

**Long, thin faces.** When the Optimize modifier is used with a high Face Thresh setting, the resulting model sometimes has long, thin faces. A face like this is much more likely to be coincident with a face on the other operand. Long, thin faces can also occur in a lofted object when the Path Steps value is set too low. You can avoid this problem by taking care when lofting objects, and by using a higher Bias setting with Optimize, such as 0.3.

**Low detail on one or both operands.** You increase the likelihood of coincident faces when there are fewer faces to work with. When preparing objects for a boolean operation, use more detail than you really seem to need. You can always bump down the face count later with the Optimize modifier.

## Tutorial 2
## Disappearing Objects

In this tutorial, you'll perform a boolean union on two simple objects. The boolean won't work at first, then you'll change the model so the boolean will work correctly.

### Step 1. Load file

Load the file BOOLTST1.MAX. This file can be found in the \MICHELE\MESHES directory on the CDROM that comes with this book.

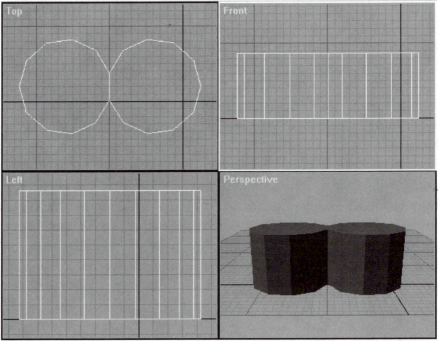

Figure 2-1. Two extruded NGons

This file contains two extruded NGons placed next to one another.

### Step 2. Boolean operation

Select one of the objects. Under the Create panel, choose Geometry. From the pulldown list, choose Compound Objects. Click on Boolean.

Under the Parameters rollout in the Operation section, turn on Union.

Click on Pick Operand B. Select the other object. Wait a few moments while MAX performs the boolean operation.

In the Perspective viewport, you can see that part of the second object has disappeared. Why did this happen?

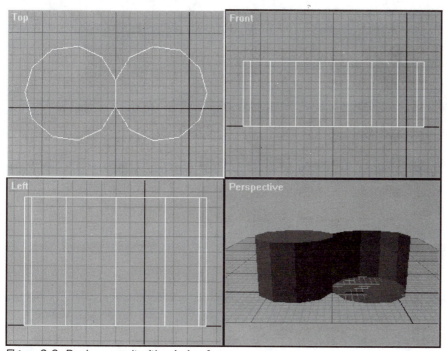

*Figure 2-2. Boolean result with missing faces*

This boolean operation failed because the objects don't firmly overlap. Where the two objects touch, the faces are occupying the same space. These coincident faces cause problems for the boolean calculation, making it impossible for MAX to figure out what's the inside and what's the outside of the boolean result.

To remedy the situation, you can move one of the objects to make the intersection more pronounced.

### Step 3. Fix boolean

*You can reload any of the last four files you used by picking it from the bottom of the File menu.*

Reload the file BOOLTST1.MAX from the \MICHELE\MESHES directory of the CDROM.

Activate the Top viewport. Click on Select and Move.  Turn on Restrict to X.

In the Top viewport, move one of the objects very slightly toward the other object. You should be able to see a slight overlap between the objects.

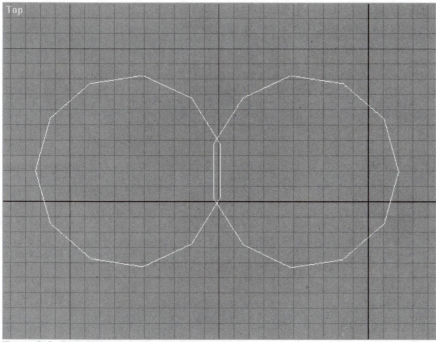

*Figure 2-3. Slightly overlapping objects*

Do the boolean operation again. The resulting object is a union of the two objects with no missing faces.

## Tutorial 3
## Stuck Objects

Sometimes a boolean subtraction goes awry and the results look more like a union. The cause, again, is coincident faces, although it's not always as obvious as it was in the last tutorial.

### Step 1. Load file

Load the file BOOLTST2.MAX. This file can be found in the \MICHELE\MESHES directory on the CDROM that comes with this book.

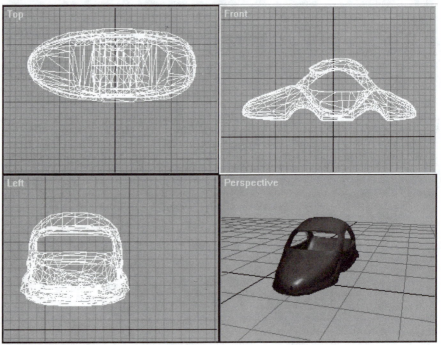

Figure 2-4. Toy car

This file contains part of a mesh of a toy car.

Suppose you want to cut away half the car to see a cross-section of the interior. The obvious approach is to make a box, position it halfway through the car and subtract it with a boolean operation. A box has already been created for this purpose.

## Step 2. Unhide box

Under the Display panel, choose Unhide All.

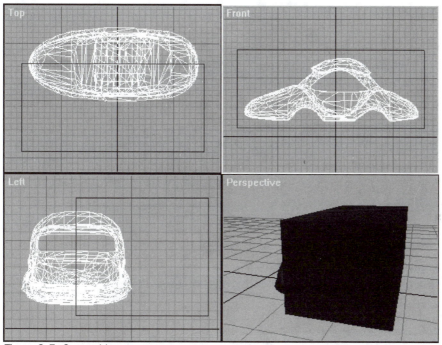

Figure 2-5. Car and box

A box appears, positioned about halfway through the car. The box will be used as a second operand to cut away half of the car.

### Step 3. Boolean operation

Next you'll perform the boolean subtraction.

Select the car. Under the Create panel, choose Geometry. From the pulldown list, choose Compound Objects. Click on Boolean.

Under the Parameters rollout in the Operation section, turn on Subtraction (A-B).

Click on Pick Operand B. Select the box. Wait a few moments while the boolean result is computed.

When the boolean operation is finished, the box is still on the screen. Click on Select and Move and try to move the box away from the car in any viewport.

Horrors! Instead of being subtracted from the car, the box has become one with the car. This isn't what we wanted at all.

The boolean didn't work because somewhere in the car are faces coincident with the face of the box that passes through the car. It's impossible to tell that this is true just from looking at the model. It took a boolean failure to tell us that.

## Step 4. Fix boolean

To remedy the problem, reload the file BOOLTST2.MAX and unhide the box again. Move the box a little to the left or right in the Left viewport. Try the boolean operation again.

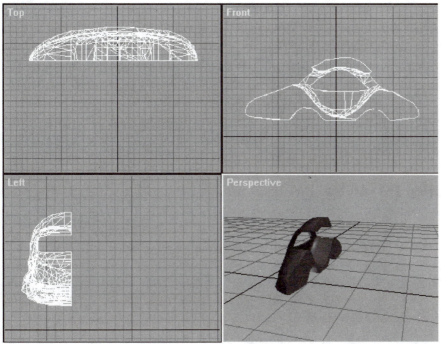

*Figure 2-6. Correctly booleaned car*

If the operation doesn't work, reload the file and try again. The boolean operation should work correctly on the second or third try.

## Tutorial 4
## Disappearing Objects

In this tutorial we'll explore some techniques for getting around the problem of coincident faces.

First you'll boolean two complex objects together with the union operation. Instead of resulting in one complete object, the two operands will disappear, a common occurrence with complex booleans. Then we'll look at different ways to solve the problem.

The solution for this particular set of operands is given at the end of the tutorial.

### Step 1. Load file

Load the file BOOLTST3.MAX from the \ALAN\MESHES directory on the CDROM.

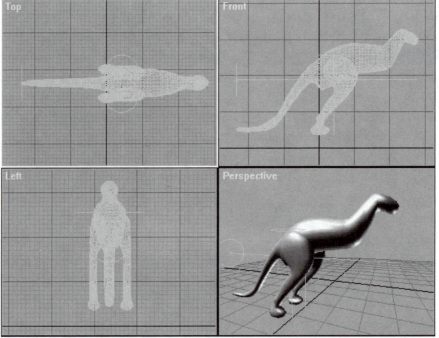

Figure 2-7. Body parts

This figure shows the beginnings of a kangaroo mesh. This mesh is part of the model you'll create in Tutorial 6.

The body and legs are three separate lofted objects. All objects have been created with medium detail. The top of each leg is embedded well inside the body, which would lead you to believe that a boolean union would work just fine.

## Step 2. Boolean union

Select the body. Under the Create panel, click Geometry, and choose Compound Objects from the list. Click Boolean. Turn on the Union button under the Parameters rollout.

Click Pick Operand B. In the Left viewport, click on the leg on the right. Everything seems to be fine. Click on the Boolean button again, click Pick Operand B and choose the other leg in the Left viewport. The mesh disappears.

*You must click the Boolean button again before you can start another boolean operartion.*

Why didn't the union work? Boolean calculations are complex, and can fail for any number of reasons. To make the boolean work, you'll have to change something about the model. That "something" can be any number of things, such as:

· **Choosing the operands in a different order**
· **Moving or rotating one or both operands**
· **Increasing path or shape steps for lofted objects**
· **Applying a modifier such as Bend or Twist to one or both operands**

It is impossible to name a surefire solution that will work with all models. The solution varies from one model to another.

### Step 3. Change operand order

Reload BOOLTST3.MAX. Do the boolean operation again, but this time select the leftmost leg in the Left viewport the first time you choose operand B. Then click Boolean again, click Pick Operand B and choose the other leg.

The boolean appears to work this time, but appearances can be deceiving.

*Using a modifier after a boolean operation is an easy, quick way to tell if the boolean was successful.*

Under the Modify panel, click on any modifier such as Bend or Edit Mesh. The object disappears.

In this case, the boolean appeared to work, but the resulting object actually isn't stable. As soon as you try to apply any modifier or use the object in another boolean operation, the mesh disappears.

## Step 4. Modify operand

Reload BOOLTST3.MAX. See if you can get the boolean operation to work on your own. Try moving, bending and rotating objects slightly.

After much trial and error, I found a solution that worked. Select the rightmost leg in the Left viewport. Go to the Modify panel and expand the Skin Parameters rollout. Change Shape Steps to 5 and Path Steps to 80. This high level of detail will help the boolean operation work properly, and can always be tempered later on with the Optimize modifier.

Under the Modify panel, choose Bend. Set the bend Angle to -10. In the Left viewport, rotate the leg -3 degress.

Try the boolean operation again with either leg as the first operand. When the boolean operation is complete, go to the Modify panel and choose any modifier. The mesh remains onscreen, which means the boolean operation was a success and the object is stable.

Remove the modifier by clicking the Remove modifier from the stack button. You can now go on to edit the object or use it in another boolean operation.

The sequence of events that worked on this particular mesh were discovered with experimentation. There is always a change that will work, you just have to find it.

Be sure to check the boolean's stability by applying a modifier before going on to another boolean operation. Save your file after each successful boolean.

## Tutorial 5
## Carving with Booleans

In this tutorial, you'll use a boolean operation to "carve" one object from another. This technique is very useful for many situations. For example, if the camera goes very close to your model, you might need detail that would require an extremely large, unwieldy texture map. Using the carving method removes the need for a texture map and preserves detail perfectly.

You also might need to create one object that fits another perfectly. In this tutorial, you'll make a perfectly fitted boot for a character's leg with the carving technique.

## Step 1. Load file

Load the file LEGNBOOT.MAX from the \ALAN\MESHES directory on the CDROM.

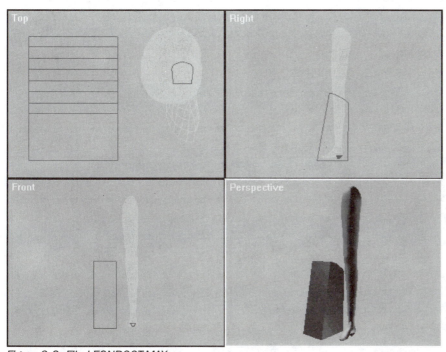

*Figure 2-8. File LEGNBOOT.MAX*

This file contains two objects, a leg and a shapeless object that approximates a boot. The leg was originally a loft object, but has been changed into an Editable Mesh. The boot object, named BootBlock, is an extruded spline. Note that the extruded spline was created with caps so it will boolean properly.

Under the Edit menu, choose Hold to hold the model in its current state.

## Step 2. Position BootBlock

Click on the Front viewport. Turn on Restrict to X. 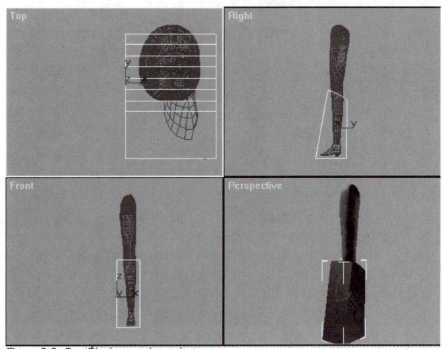 Move
BootBlock to the right to sit over the bottom part of the leg.

*Figure 2-9. BootBlock moved over leg*

Look in the Front and Left viewports to make sure the bottom of
the leg is completely enclosed in BootBlock.

Under the Edit menu, choose Hold to hold the model in its current state.

### Step 3. Boolean operation

Select BootBlock. Under the Create panel, click Geometry and choose Compound Objects from the pulldown list. Click on Boolean.

Under Operation, pick Intersection. Click on Pick Operand B. Pick the leg object.

The result is the intersection of the two objects resembling a high boot.

### Step 4. Save boot separately

Give the new boot the name Boot. With Boot selected, choose Save Selected from the File menu. Enter the name BOOT.MAX and save the file to your hard disk.

### Step 5. Create new leg

Under the Edit menu, choose Fetch to bring back the original model.

Select the leg. Under the Create panel, click Geometry and choose Compound Objects from the pulldown list. Click on Boolean.

Under Operation, pick Subtraction (A-B). Click on Pick Operand B. Pick BootBlock.

The result is a leg that terminates cleanly where the BootBlock object ended.

### Step 6. Merge boot

Under the File menu, choose Merge. Choose the file BOOT.MAX and pick the object Boot to merge.

You now have two completely different objects that fit together perfectly. If you like, you can enlarge the boot slightly and add trim, buckles or wrinkles to make the boot look more real.

### Things to Try

This carving technique can be used in many different ways.

*Figure 2-10. Pod*

Figure 2-10 shows a pod object. The model specifications called for a number five to be painted on the pod. One way to do this is to use a texture map with the number five on it and position the UVW coordinates. However, this could take a lot of tweaking, and the resulting image wouldn't stand up to close-ups. Instead, an extruded spline in the shape of the number five was carved from the pod, then merged back into the file.

Figure 2-11. Pod with carved five

The number five is a separate object, but it looks as though it were painted onto the pod.

This carving technique has many more uses. If you have a model of rough terrain, for example, you can carve out a road that follows the terrain. You can try out this technique on the terrain model you'll make in Tutorial 18.

# Tutorial 6
# Animated Boolean Tunnel

A loft object can be animated and booleaned, so why not an animated boolean? The technique described in this tutorial uses an animated tunnel as a boolean operand. In this way, the tunnel is drilled through an object over the course of the animation.

This technique is easy to implement and can be used for many different effects. A camera and a light traveling through the tunnel just behind the opening, along with an interesting texture map on the object being drilled, can make a great opening sequence for a demo reel.

## Step 1. Create objects

Under the Create panel, click Geometry. Click Box. Create a box in any viewport. Under the Parameters rollout, change the Length, Width and Height values to 100.

Click Zoom Extents All, then click Zoom All. Drag down in any viewport to create some room around the box. Click Select and Move to get out of zoom mode.

Click Shapes. Click Circle. Create a small circle in any viewport.

Under the Parameters rollout, change Radius to 5.

Next you'll create a loft path. The path will be used to create a tunnel to be booleaned through the box.

Click Line. In the Top viewport, draw a curved line starting outside the box, passing through the box and coming out the other side, as shown in Figure 2-12.

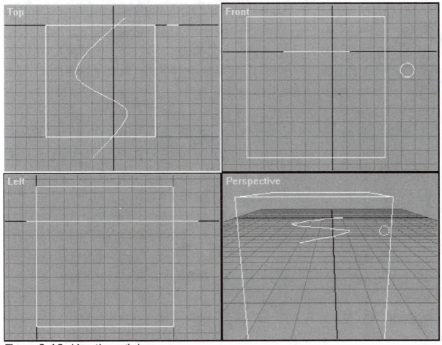

*Figure 2-12. Line through box*

Name the line PathLine.

Under the Modify panel, click Edit Spline. Click Select and Move.

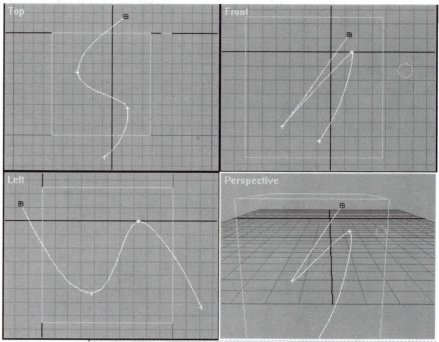 Move vertices on the line in all viewports to make the line pass through the box in all three dimensions. As necessary, change the vertex types to Bezier and adjust the handles to make a smooth curve. See Figure 2-13.

Figure 2-13. Three dimensional line

Under the Create panel, click Geometry. Choose Loft Object from the pulldown list. With PathLine still selected, click Loft. Click Get Shape, and click on the small circle. Under the Skin Parameters rollout, turn on the Skin checkbox. Increase Path Steps to 10 to make a smooth loft object.

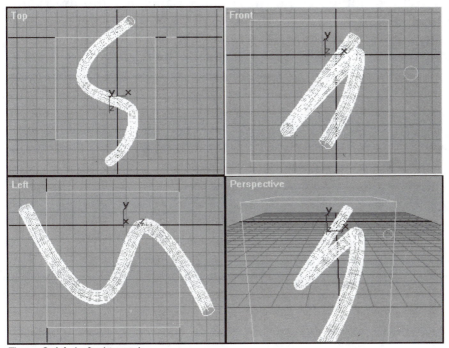

Figure 2-14. Lofted tunnel

Name the object Tunnel.

### Step 2. Set up animated scale

This tunnel will eventually be used as a boolean operand to carve a tunnel through the box. First you'll animate the scale of the lofted object so the tunnel grows over time.

Select the tunnel. Under the Modify panel, expand the Deformations rollout and click Scale. The Scale Deformation window appears.

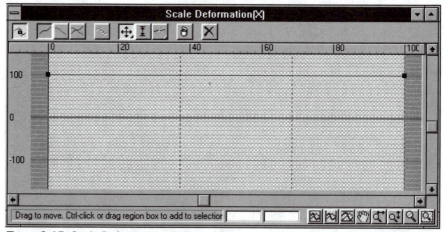

Figure 2-15. Scale Deformation window

Click Insert Corner Point. 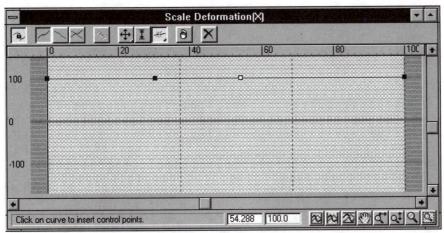 Click on the red scale line at any two spots to create two more control points.

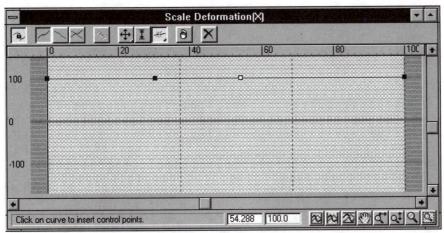

*Figure 2-16. Two new control points*

Click Move Control Point. Click on the second control point. The point turns white to indicate that it is selected. Enter 50 in the leftmost box at the bottom of the Scale Deformation window.

Click on the third control point. Enter 50.01 in the leftmost box, and in the rightmost box type 0.01. These two control point values will be used to animate the size of the loft.

*It is very important that you set the scale percentage to 0.01% and not 0%. A 0% setting will cause the animated boolean to fail.*

Click on the control point at the far right. Change the rightmost text box from 100.0 to 0.01. The window should look like Figure 2-17.

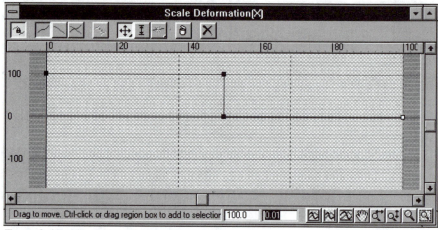

*Figure 2-17. Adjusted control points*

The leftmost box represents the location on the path as a percentage. The rightmost box is the scale percentage of the loft object's diameter. Right now, the loft object is set up to start out at 100% of its original size and stay that way halfway up the path. At that point, the path diameter is scaled to 0.01% of its original size and remains that way until the end of the path.

These control points will be used to animate the loft object's diameter along the path.

## Step 3. Animate tunnel scale

Draw a bounding box around the two middle control points to select them. Drag them all the way to the left to frame 0. The loft object scales to 0.01% of its original size.

Click the Animate button to turn it on. Drag the frame slider to 100. Drag the control points all the way to the right to frame 100. The tunnel scales up to its original size. Close the Scale Deformation window.

Drag the frame slider back and forth. The Tunnel object grows over time.

Under the Modify panel, click on More and choose Material from the list. Under the Parameters rollout, change Material ID to 2.

## Step 4. Animated boolean

The animated tunnel can now be used to carve a growing tunnel through the box.

Move the frame slider so part of the tunnel is visible. Select the box. Under the Create panel, click Geometry. Choose Compound Objects from the pulldown list.

Click Boolean. Make sure Subtraction (A-B) is selected. Click the Pick Operand B button and click on Tunnel.

Select Tunnel. The easiest way to do this is to click on the tunnel outside the box. Under the Display panel, choose Hide Selected.

You now have an animated tunnel that carves its way through the box. Move the frame slider back and forth to see the animated boolean.

### Step 5. Camera and light

Turn off the Animate button.

Under the Create panel, click Lights. Create an omni light in any viewport with the default settings.

Under Create, click Cameras. Click on Free and place the camera in the Front viewport. Under the Parameters rollout, choose a Stock Lens of 28mm and make the Target Distance 25 units. Change the lower right viewport to the camera view. To do this, click on the Perspective viewport and press the <C> key.

Go to frame 0. Select the camera. Under the Motion panel, expand the Assign Controller rollout. Highlight Position:Bezier

Position. Click the Assign Controller button. Choose Path from the list.

Under the Path Parameters rollout, click Pick Path. Select the path in any viewport. Turn on the Follow checkbox.

The camera has moved to the start of the path, but it's pointing in the wrong direction. Rotate the camera in one or more viewports so it points in the direction of the path.

Select the omni light. Use the same procedure to make the light follow the path.

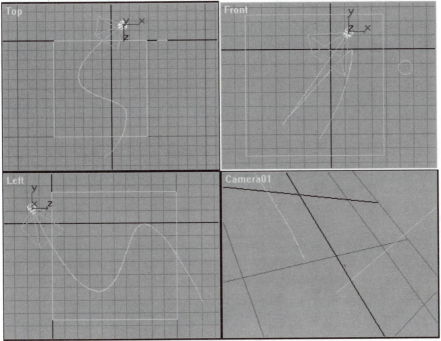

*Figure 2-18. Camera and light at start of path*

## Step 6. Offset camera and light

The camera and light currently move with the tunnel object, and are essentially right on top of the opening of the tunnel at all times. In order to see the tunnel as it opens, the camera and light must be just behind the opening. You'll use Track View to make this happen.

Click the Track View button. Expand the Objects listing to see the tracks for the light and camera. Click and drag on the camera track to move it 5 frames to the right. This will make the camera travel just behind the opening of the tunnel. Drag the light track 5 frames to the right also.

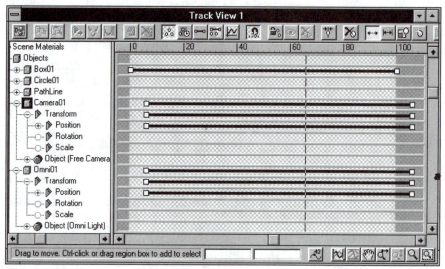

*Figure 2-19. Camera and light tracks offset*

Close Track View.

Turn on the Animate button. Go to frame 20. If necessary, rotate the camera so it looks right down the tunnel. Repeat for frames 40, 60, 80 and 100.

The camera view flies the viewer through the tunnel. Render the camera view to an AVI file.

To really see the tunnel effect, create a checker material and apply it to the box. Be sure to apply a UVW modifier to the box.

To see an animated sequence made with this model, look at the file TUNNEL.AVI in the \TED\AVI directory on the CDROM.

# c h a p t e r

## Kangaroo

In this chapter you'll create and animate a kangaroo. The kangaroo will be built with many tools including fit deformation. To make the kangaroo jump, the Character Studio plug-in will be used.

*Figure 3-1. Kangaroo*

In MAX, several tools are needed to make a model of an animal. When you set out to create an animal, you must first break its body into parts and decide on the best method to make each one.

To make the kangaroo, you'll break it down into several parts.

**Body**
**Front legs**
**Back legs**
**Fingers**
**Toes**
**Ears**
**Eyes**

The body and legs will be created with fit deformation. The fingers and toes will be made from cylinders, the ears from tubes and the eyes from spheres.

Fit deformation can be confusing if you've never used it. Before we begin the tutorial, here is a brief description of how fit deformation works.

## How Fit Deformation Works

Fit deformation is a modeling technique that works with profiles. You provide the top and side profiles for an object, and fit deformation creates a 3D object from these profiles.

For example, suppose you want to make a car body. From the side, a car body looks something like the shape below.

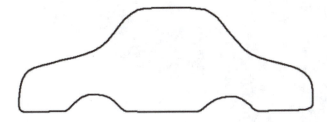

Figure 3-2. Car side profile

From the top, the car looks like Figure 3-3.

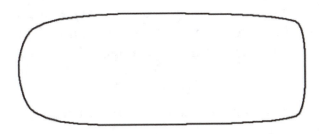

Figure 3-3. Car top profile

These two shapes can be used to make a car body with fit deformation.

*Figure 3-4. Car body created with fit deformation*

A third shape is used with fit deformation to determine how the two profiles will be put together. In most cases, a circle or rectangle will work fine. The size of this third shape is not important, but whether its edges are rounded or angled does make a difference.

In Figure 3-4, a rounded rectangle, such as the one in Figure 3-5, was used as the third shape.

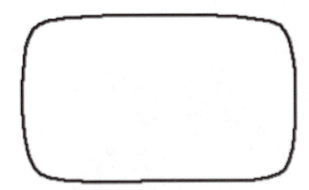

*Figure 3-5. Third shape for fit deformation*

In Figure 3-6, a circle was used as the third shape for the car.

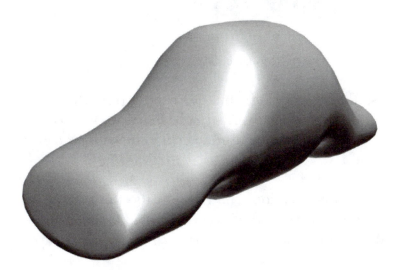

*Figure 3-6. Car body created with circle*

A rectangle was used to create the car body in Figure 3-7.

*Figure 3-7. Car body created with rectangle*

To use fit deformation, you also need a path. In most cases, a straight line works best. A straight line path was used to create the car, and will also be used to create the kangaroo body in the next tutorial.

To use fit deformation, you first loft the third shape along the path. You then go to the Modify panel and choose Fit from the Deformations rollout. The Fit Deformation window appears.

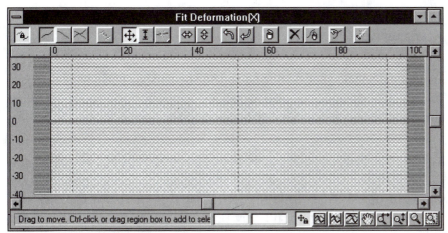

*Figure 3-8. Fit Deformation window*

From this window you can select the top and side profile shapes. MAX uses this information to create the 3D object.

Even if you're not completely clear on how fit deformation works, do each step of the tutorial that follows. After each step you'll see the effect that each change makes on the object, and the concepts of fit deformation will become clearer.

## Tutorial 7
## Kangaroo

In this tutorial, you'll build a kangaroo torso with fit deformation. You'll draw a top and side view of each body part, and also supply a third shape for lofting along the path.

When creating a model of an animal, it is essential that you use a photograph, drawing or sculpture of the animal as a reference. A drawing of a kangaroo is provided on the CDROM for use as a guide when making each body part.

The body is lofted from a group of splines that you create over photos of a kangaroo. Start by creating a spline which follows the contours of the kangaroo's body.

### Step 1. Set up viewports

First you'll set up the Top viewport for modeling.

Activate the Top viewport. Click the Min/Max Toggle button  to maximize the Top viewport.

From the Views menu, choose Background Image, then click on Files. Load the file KANGBGD.BMP from the CDROM that comes with this book. This file can be found under the \SANFORD\MAPS directory.

Right-click on the Top viewport name. Turn on Show Background. A side view of a kangaroo appears in the viewport.

*Fit deformation is easiest to use when the Top viewport is used for spline creation.*

*Figure 3-9. Kangaroo background*

## Step 2. Draw side profile

First you'll create the shapes to be used in the deformation.

Under the Create panel, click on Shapes. Click on Line.

Draw an outline around the kangaroo body shape in the background image. Don't include the ears, front legs or rear legs. Make the spline look as much as possible as Figure 3-10, but don't be overly concerned if it's not exactly right. You'll correct the spline in the steps that follow.

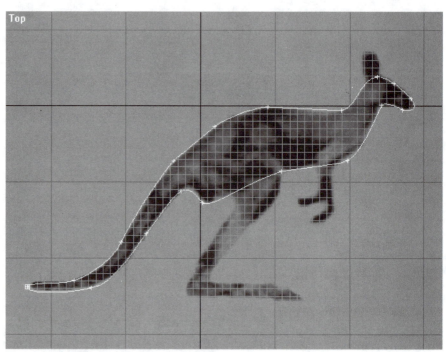

*Figure 3-10. Spline around kangaroo body*

Use about 16 vertices to define the shape. Draw all the way around and place the last point over the first point. Close the spline.

## Step 3. Modify profile

Select the line and go to the Modify panel. Click on Edit Spline. In the Top viewport, move the points on the spline until they fit the outline closely.

If the curve on some parts of the line is inaccurate, choose Select and Move. Right-click on the vertex and change the type to Bezier if necessary. Adjust the Bezier handles until the curve fits the kangaroo body shape.

Once you've drawn the shape, you no longer need to see the background. To remove the background image from the screen, right-click on the Top viewport name. Turn off Show Background.

*If you don't know how to adjust Bezier handles, see Practice Exercise 4 in Chapter 1.*

### Step 4. Modify shape for fit deformation

Fit deformation shapes work best when there is a vertex at each end. This means that there should be a vertex at the leftmost and rightmost part of the shape when viewed in the Top viewport. Compare the two shapes below.

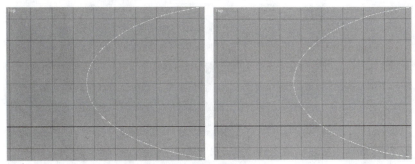

*Figure 3-11. No vertex at the end, vertex at the end*

In Figure 3-11, the leftmost end of the left shape has no vertex. The shape on the left is set up for fit deformation, with a vertex at the leftmost position.

If you don't have vertices at the ends of the kangaroo body shape, you can use Refine to put vertices there.

In the Top viewport, use Zoom Window 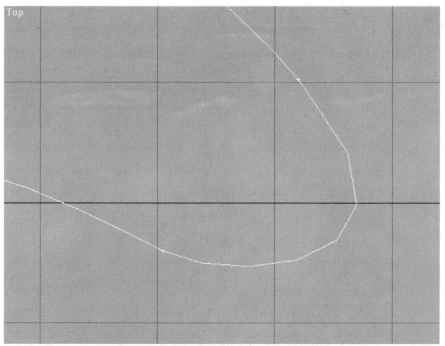 to zoom in on the rightmost end of the kangaroo shape.

Figure 3-12. Zoom on right end of shape

You should still be in the Edit Spline modifier with Sub-Object turned on.

Under the Edit Vertex rollout, click on Refine. When you move the cursor over the shape, it changes to include two short lines. Move the cursor to the rightmost part of the shape, and click to create a new vertex.

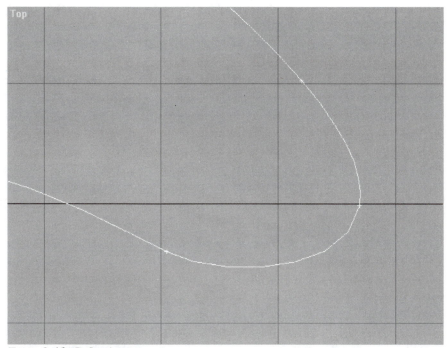

*Figure 3-13. Refined vertex*

Right-click on the new vertex, and change the type to Bezier if necessary. Right-click again and choose Move from the pulldown list.

*Choosing the Move option from the pull-down list has the same effect as clicking on Select and Move.*

Move the Bezier handles so they point straight up and down, as shown in Figure 3-14.

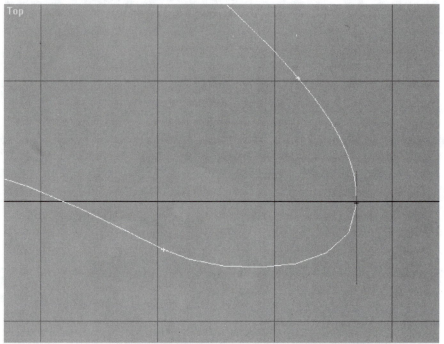

Figure 3-14. Bezier splines up and down

Click Zoom Extents.

Click Zoom Window to zoom in on the leftmost end of the spline. Click Refine, and place a vertex at the leftmost part of the shape. Change the vertex type to Bezier, and edit the handles so they point straight up and down.

The side profile is complete. Name the spline SideBody.

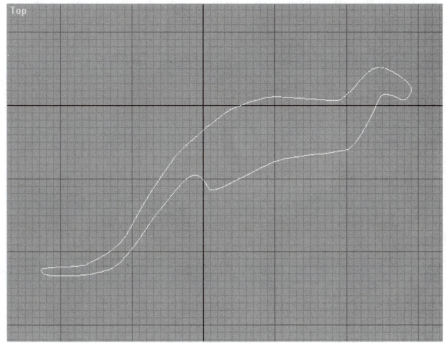

Figure 3-15. Spline around kangaroo body

### Step 5. Create top profile

Next you'll create a spline to represent the top profile of the kangaroo.

Under the Create menu, click on Shapes and choose Line. Under the Creation Method rollout, turn on Smooth for Initial Type. In the Top viewport, create the spline shown below. Don't be concerned about making the spline look exactly like the shape below. Just make it as close as you can. Make the spline the same length as the side view of the kangaroo.

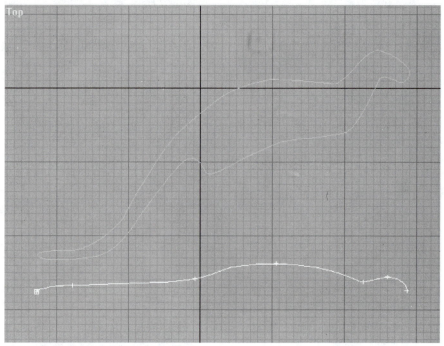

*Figure 3-16. Top profile kangaroo body*

Under the Modify menu, click Edit Spline, and click on Sub-Object under the Modifier Stack rollout. Adjust the vertices as necessary in the same way you adjusted the vertices for the side view of the kangaroo body.

Make sure the vertices on each end of the spline are on the same horizontal line. This will help you mirror the spline. You can select and move the spline to sit on a grid line to check if the end vertices are on the same line. Move one or both end vertices as necessary.

On the Modifier Stack rollout, for Sub-Object, choose Spline. Click on the spline to select it. The spline turns red.

Look under the Edit Spline rollout. Next to the Mirror button, turn on the Copy checkbox. Under the Mirror button, turn on

Mirror Vertically. ⬓ Click on Mirror. The shape is mirrored and copied vertically.

Choose Select and Move. ✛ Move the one shape so the ends of the two shapes touch. A dialog box appears.

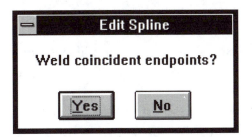

Click on Yes to weld the ends together.

> **TIP** *The method used to create this shape ensures that there will be a vertex at the leftmost and rightmost ends.*

The top profile is now complete. Name the spline TopBody.

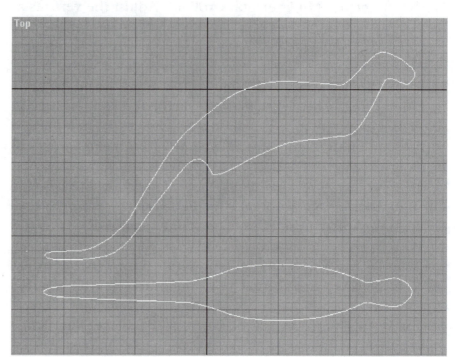

*Figure 3-17. Side and top profiles*

Check the new top view shape against the side view shape to make sure they correspond. For example, the neck on both shapes should begin to taper at the same place, as should the tail. To check this, draw an imaginary vertical line through the top view where the neck begins to taper. Where the imaginary line passes through the side shape, the neck should begin to taper also. Check the tail in the same way.

If it is necessary to adjust vertices, click on Sub-Object under the Modifier Stack rollout, and make sure Vertex is turned on. Move each vertex as necessary to make the shapes correspond.

## Step 6. Create cross section shape

Fit Deformation requires three shapes: a top view, a side view, and a third shape to be lofted along the path. To figure out what this shape should look like, imagine that you have a sculpture of a kangaroo. If you cut the kangaroo body in half vertically, ignoring its arms and legs, the shape would be roughly a tall ellipse. The same can be said for the tail and head.

This third shape will be squashed and stretched to fit the top and side shapes. For this reason, a circle will work fine for the third shape.

Create a circle between the top and side shapes. Name the shape CrossBody.

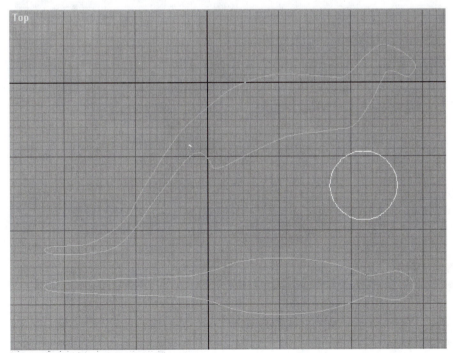

*Figure 3-18. Circle cross section*

### Step 7. Create path

Now that the three shapes have been created, you need a path for lofting.

Under the Create menu, choose Shapes and click on the Line button.

Draw a single line across the Top viewport, the same length as the top and side views. Use only two vertices to draw the line, one to start the line and one to finish. Make the line straight across. Name the line PathLine.

**When making a straight path for lofting, always make all vertices Corner type.**

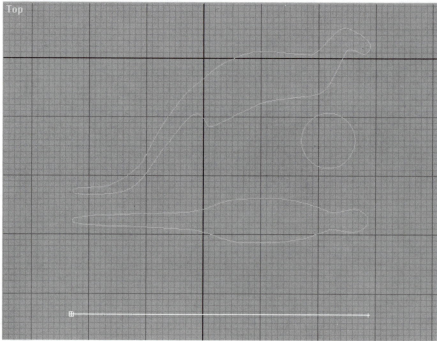

*Figure 3-19. Path line*

Under the Modify panel, choose Edit Spline. Turn on Sub-Object and choose Vertex as the sub-object level. Right-click on the leftmost path vertex and change it to Corner type if necessary. Do the same for the rightmost vertex.

Click Refine, then add the five vertices to the path. After you create each vertex, right-click on it and change the type to Corner if necessary. Turn off Sub-Object.

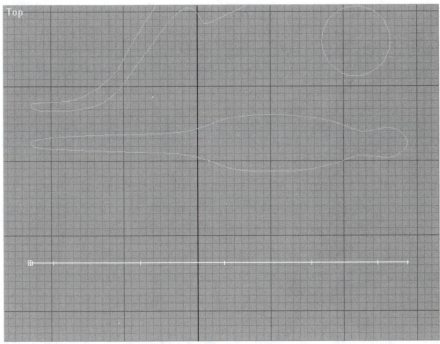

Figure 3-20. Path with five vertices added

## Step 8. Loft

Now you'll put all the shapes together to make the kangaroo body. First you'll loft the circle.

Click Min/Max Toggle to return to the four-viewport layout.

*Putting a vertex near each end, then adding a few vertices at regular intervals along the middle of the path, will make smoother fit deformation objects.*

Select PathLine. Under the Create menu, click on Geometry. On the pulldown menu, choose Loft Object.

Click on the Loft button. The Loft rollout appears.

Under the Creation Method rollout, click on the Get Shape button. In the Top viewport, click on the circle. A copy of the circle moves to the end of the path.

You have just lofted the circle, but you can't see the lofted object yet. Under the Skin Parameters rollout, turn on Skin under the Display heading. The mesh of the lofted ellipse appears in all viewports.

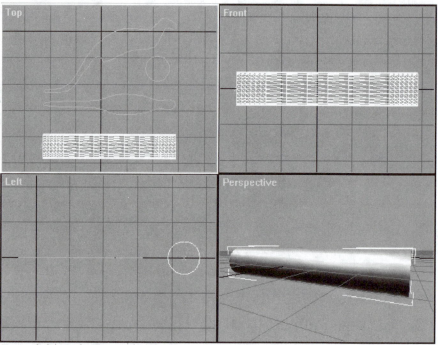

*Figure 3-21. Lofted mesh*

Name the loft obejct Body.

## Step 9. Fit deformation

Select the lofted cylinder. Click on Modify. Locate the Deformations rollout at the very bottom of the command panel. Click on the Fit button. The Fit Deformation window appears.

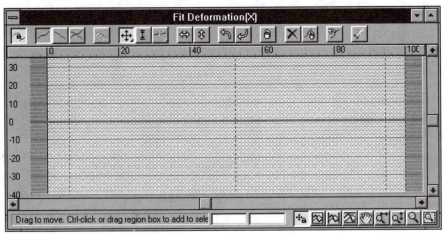

Figure 3-22. Fit Deformation window

Drag the window so it covers the bottom half of the screen, but don't move it so far that you can't access all the buttons. This will give you room to select shapes and see the deformation as it happens.

Turn off the Make Symmetrical button. ![button] This will allow you to bring in two different shapes for the top and side profiles.

By default, when the Fit Deformation window appears it's ready for the X profile. Click on Get Shape, ![button] and click on the top profile spline Topbody in any viewport. The profile appears in the window.

**TIP** *In Fit Deforma-
tion, the terms
X and Y are
used to describe the
two different profiles
you'll be using. These
X and Y terms have
nothing to do with the
X and Y axes used to
create and scale ob-
jects, nor with any
other X and Y terms
used anywhere else in
MAX.*

To see the entire profile, click the Zoom Extents button 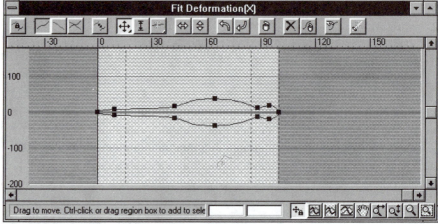 at the bottom of the Fit Deformation window.

*Figure 3-23. Top profile in Fit Deformation window*

The lofted object has also changed to reflect this new shape.

Click on Display Y axis. The X profile disappears. The Fit Deformation window is now prepared to accept the Y shape.

The Get Shape button is still pressed, so there's no need to press it again. Click on SideBody in any viewport. The shape appears in the window.

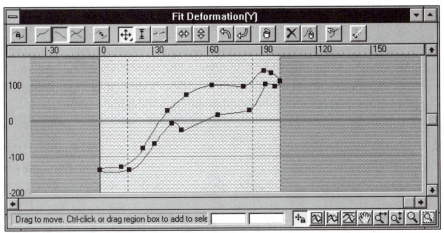

*Figure 3-24. Side profile in Fit Deformation window*

Close the Fit Deformation window. Look at the kangaroo body in all viewports. It should look similar to Figure 3-25.

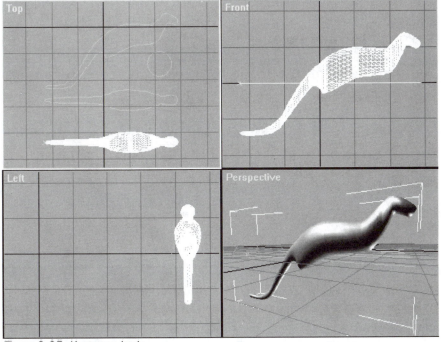

*Figure 3-25. Kangaroo body*

> **TIP** *If your kanga-roo looks radi-cally different from Figure 3-25, check over the steps to see where you may have gone wrong. If you made your shapes in a viewport other than the Top viewport, the fit shapes may have come into the Fit Deformation window at the wrong orienta-tion. In the Fit Defor-mation window, check that the X and Y shapes look like Fig-ures 3-23 and 3-24. If not, you can try press-ing the Rotate 90 CCW button repeatedly un-til the shapes are in the same orientation as those in the figures. If this happens, you will also have to rotate the kangaroo body af-ter it's lofted to make it look like Figure 3-25.*

### Step 10. Path steps

The kangaroo body mesh has a lot of vertices in some places, and not so many in others. Right now, the distribution of vertices is determined by the top and side profiles. The object detail would be better determined by the vertices on the path.

Under the Skin Parameters rollout, turn off the Adaptive Path Steps checkbox. Note that the path steps are now controlled by the path. The kangaroo body, however, looks boxy.

*The higher you set the number of Path Steps, the longer it will take MAX to redraw the screen and render the model. Set Path Steps as low as you can while retaining model smoothness.*

Increase Path Steps to 16. The kangaroo looks smoother. The nose is smooth and round, and there are considerably fewer vertices in the mesh.

If your redraw time slows down considerably after you increase the number of path steps, lower it to 8 or 10.

### Step 11. Hide and save

You don't need all the shapes onscreen any more, so you can hide them to clear up some workspace.

*The path and shape for the kangaroo body are an integral part of the lofted object. If you delete or hide the path or the shape on the path, the kangaroo body will disappear too.*

Select SideBody, TopBody and PathLine. Don't select the circle, the kangaroo body, or the shape on the path. On the Display panel under the Hide by Selection rollout, choose Hide Selected.

Save your work as KANG01.MAX.

### Step 12. Leg shapes

Activate the Front viewport. Display the background again by right-clicking on the Front viewport label and turning on Show

Background. Click Min/Max Toggle.

Use Zoom [image of magnifier icon] and Pan [image of hand icon] to move the kangaroo body so it fits right over the picture.

Under the Create panel, click Shapes, then Line. Create splines for the back leg profiles, as shown in Figure 3-26. Also create a straight-line path the same height as the shapes. The procedure for making the leg shapes is similar to the process you used for creating the shapes for the kangaroo body.

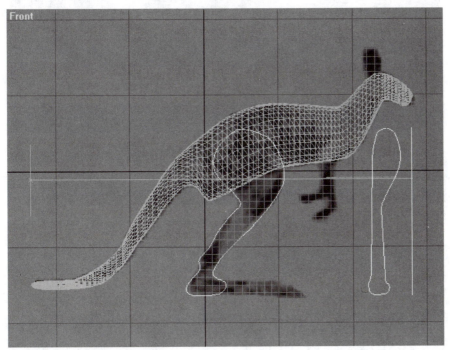

Figure 3-26. Back leg shapes

These profiles don't include the kangaroo's toes. Those will be created separately later on.

Now create three more shapes for the front leg, as shown in Figure 3-27.

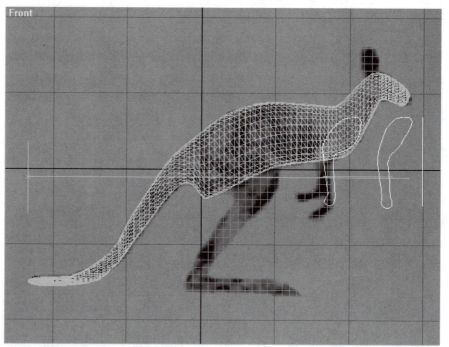

Figure 3-27. Front leg shapes

These shapes will be rotated by 90 degrees during the fit deformation process. For this reason, it is important that you put vertices at the very top and bottom of the shape, rather than the extreme right and left.

You no longer need the background. Right-click on the Front viewport label and turn off Show Background.

## Step 13. Loft back leg

Click Min/Max Toggle [icon] to return to the four-viewport display. Select the path for the back leg. Under the Create menu, click on Geometry and choose Loft Object from the pulldown menu. Click on Loft, then click on Get Shape. Pick the circle. Under the Skin Parameters menu, turn on the Skin checkbox.

With the skinned object still selected, go to the Modify panel. Under the Deformations rollout, choose Fit.

Turn off Make Symmetrical. [icon] Choose Get Shape, [icon] and pick the thinner shape you made for the back leg. Click the Rotate 90 CCW button [icon] to orient the shape.

Click on Display Y Axis, [icon] and pick the wider back leg shape.

Click the Rotate 90 CCW button. [icon] The back leg mesh appears in the viewports. Close the Fit Deformation window.

Under the Skin Parameters rollout, turn off Adaptive Path Steps. Change Path Steps to 36.

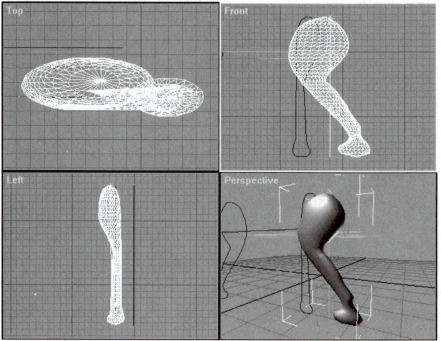

*Figure 3-28. Back leg*

Name the leg BackLegL.

Turn on Angle Snap Toggle. Rotate and move the leg to sit against the kangaroo body. The top of the leg should be inside the kangaroo body.

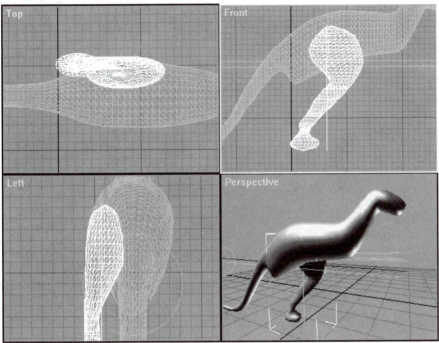

*Figure 3-29. Back leg position*

### Step 15. Loft front leg

Select the path for the front leg, and loft the circle along it. With the same technique you used for the back legs, apply a fit deformation to the loft object. Use the thin shape as the X shape and

the wider shape as the Y shape. Click Rotate 90 CCW 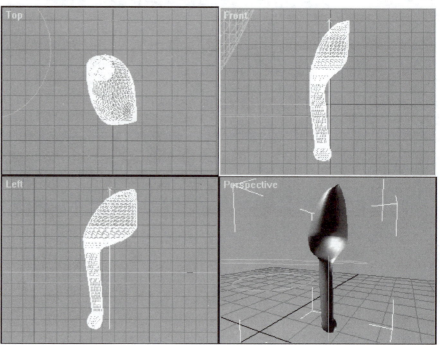 after you bring in each shape.

Under the Skin Parameters rollout, turn off Adaptive Path Steps. Change Path Steps to 16.

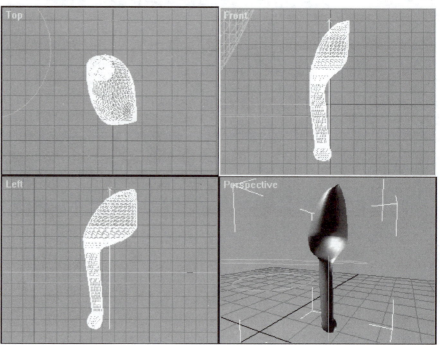

Figure 3-30. Front leg

Name the leg FrontLegL.

Rotate and move the leg to the correct position against the body. The top of the leg should be firmly inside the body.

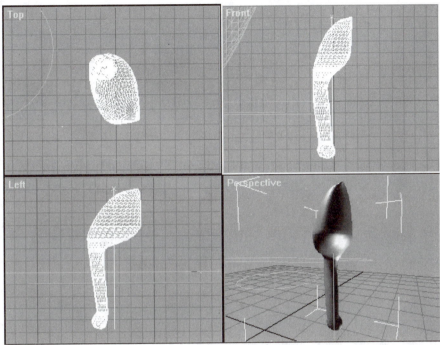

Figure 3-31. Front leg position

Select all the shapes used to make the front and back legs. Under the Display panel, choose Hide Selected.

### Step 17. Mirror legs

Next you'll mirror the legs to make a set of right legs.

Select the front and back legs. Activate the Left viewport.

Click the Mirror Selected Objects button 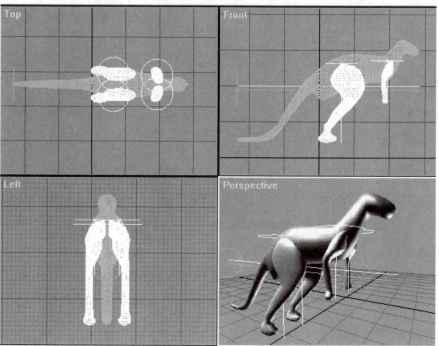 on the toolbar. Under Clone Selection, turn on Copy. Click OK to mirror the legs.

Move the new legs to the other side of the kangaroo body. Name the new legs BackLegR and FrontLegR.

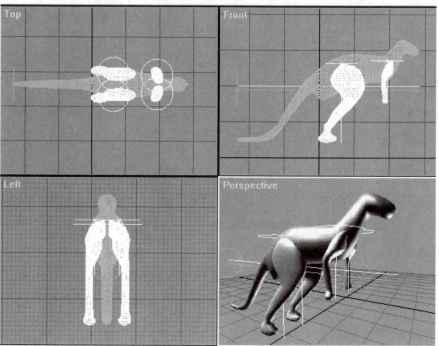

*Figure 3-32. Body parts positioned*

Save your work as KANG02.MAX.

**Step 18. Assemble kangaroo**

Now you'll use the boolean union operation to put together the kangaroo.

Select the kangaroo body. Under the Create panel, click Geometry. Choose Compound Objects from the pulldown list.

Click Boolean. Under the Parameters rollout, turn on Union. Click Pick Operand B. Choose one of the legs.

After a few moments, the leg and body will appear as one object.

Save the file as KANG03.MAX.

Click the Boolean button again, and union another leg to the body. After each successful boolean, save the file with a new filename. This will allow you to reload the file if the boolean doesn't work.

When you're finished unioning all the legs onto the body, change the object name to Body if necessary.

Save the file as KANGB01.MAX.

*A boolean union sometimes results in either or both objects disappearing. If this happens, restore the previously saved file, move or rotate the leg slightly in any direction, save the file and try the boolean operation again.*

*If you're still having trouble getting the booleans to work, see Tutorial 4 for more information on dealing with troublesome booleans.*

*Be sure to click the Boolean button before each boolean operation. If you don't, MAX will think you're picking a new operand B for the previous boolean operation.*

*Optimization can create long, thin faces that cause rendering artifacts. A higher Bias setting such as 0.3 keeps the optimized faces from being too long and thin.*

## Step 19. Optimize mesh

Next you'll optimize the mesh to remove extra vertices in areas where they're not needed. This will speed up redraw time and make it easier to animate the model.

Select the body. Under the Modify panel, click on More and choose Optimize from the list. MAX begins the optimization process immediately. It may take a few moments to calculate the new mesh.

Under the Parameters rollout, change Face Thresh to 3.0 and Bias to 0.3.

To improve redraw time, collapse the modifier stack. Click on

Edit Stack. Choose Collapse All on the dialog box. Click OK to exit.

Save the file as KANGB02.MAX.

## Step 20. Smooth joints

The legs and body are now all one object, but there are clear seams where the legs meet the body. To smooth out the joins, you'll use the MeshSmooth modifier on selected parts of the mesh. MeshSmooth adds faces and relaxes angles between those faces to smooth out sharp corners.

Select the kangaroo body. Under the Modify panel, choose Edit Mesh. Choose Face as the Sub-Object level.

In the Front viewport, zoom in on where the back legs meet the body. Select faces where the front part of the legs join the body.

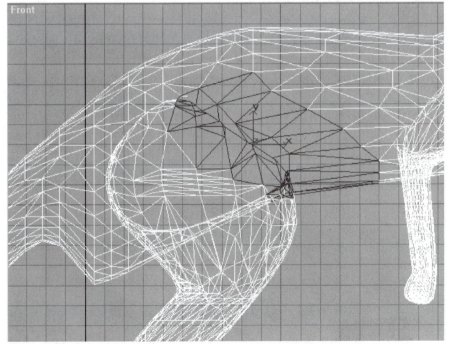

Figure 3-33. Selected faces

*The sequence described in this step — select faces, apply Optimize, select faces, apply MeshSmooth — can be used to smooth the joins on any boolean union.*

Under the Modify panel, click More. Choose Optimize from the list. Only the selected faces will be optimized.

You won't see any difference in the Perspective viewport just yet. Look only at the Front viewport to see if the faces look as though they're flattening out. If not, try a higher Face Thresh value such as 10.0 or 16.0.

Choose Edit Mesh and choose the Face level for Sub-Object. Select the faces in the same area once again.

Click More again and choose MeshSmooth from the list. MeshSmooth takes a few moments to calculate the new mesh based on the default parameters.

Set Strength, Relax and Sharpness to 0.5, and turn on the Smooth Result checkbox. When MeshSmooth is finished, the join at the front of the leg should look much smoother in the Perspective viewport.

*Figure 3-34. Smoothed joint*

Don't be concerned about slight puckering around the smoothed area. This problem will be taken care of when you smooth the entire object later in this step.

Repeat this process for the front legs. When you've finished smoothing the joints, collapse the stack.

Next you'll smooth the entire object.

Click More, and choose MeshSmooth. Set Strength, Relax and Sharpness to 0.5, and turn on the Smooth Result checkbox.

If you want the mesh smoother, you can increase Iterations to 2.

Collapse the stack, and save the file as KANGB03.MAX.

## Step 21. Toes

Next you'll make the fingers and toes for the kangaroo.

Under the Create panel, click Geometry and choose Cylinder. In the Front viewport, make a tall cylinder about ¼ the length of the kangaroo's body. In the Parameters rollout, set Height Segments to 12 and the Cap Segments at 5.

*If you choose Edit Mesh and select faces or vertices, then pick another modifier, the new modifier is applied to the selected faces or vertices only. An asterisk before a modifier name in the stack indicates that the modifier was applied to just the selected faces or vertices.*

*TIP* **If your system resources are running low or the screen is taking a long time to redraw, start a new file for the fingers and toes. Later you can merge the file with KANGB03.MAX.**

*TIP* **A bitmap file showing the toe can be loaded as a background for the Left viewport. Using this file as guide, you can easily move and rotate the vertices to match the picture of the toe. Load the file KANGTOE.BMP from the \SANFORD\MAPS directory of the CDROM.**

Under the Modify panel, choose Edit Mesh. In the Left viewport, select different columns of vertices. Move and scale the vertices until the toe looks similar to Figure 3-35.

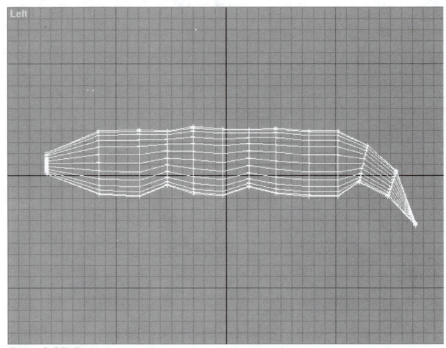

Figure 3-35. Toe

Once you have one toe created, create a copy and set it aside. This object will be used to create the fingers.

Click Select and Non-uniform Scale. Activate the Top viewport and turn on Restrict to Y. Scale the toe to about 200%.

Make two copies of the toe. Scale one toe to be slightly longer than the others. Rotate and position the toes against the right foot, as shown below. You can reload the background image KANGBGD.BMP as a guide for placing the feet.

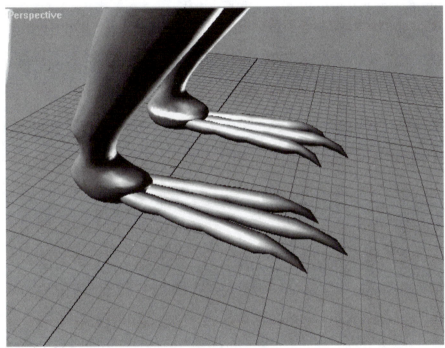

Figure 3-36. Positioned toes

Select all the toes and link them to the kangaroo body with Se-

lect and Link.

Save the file as KANGB04.MAX.

### Step 22. Fingers

A kangaroo has five fingers on each of its front legs.

To create a finger, click Select and Uniform Scale. 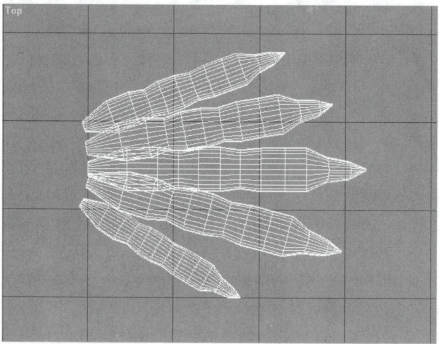 Scale the finger copy down to about 50% of its original size. Move the finger near the right front leg.

Copy the finger to make five fingers. Move, scale and rotate the fingers as shown in Figure 3-38.

*Figure 3-37. Finger positions*

Position the fingers against the ball of one of the front legs, using the background image KANGBGD.BMP as a guide.

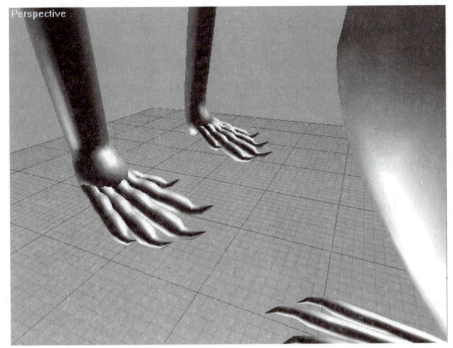

*Figure 3-38. Finger positions*

Select all the toes. Click Select and Link. 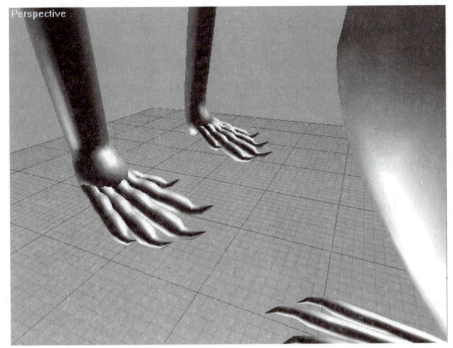 Click and drag to link the toes to the body.

Save the file as KANGB05.MAX.

> **TIP** *The joins on the hands and feet are only visible if you zoom in on them, so the seams are not much of a concern.*

### Step 23. Ears

The ears are made from simple tubes with only one height segment. To make sure you don't accidentally modify the body mesh, select the body. Under the Display panel, choose Freeze Selected under the Freeze by Selection rollout.

Under the Create panel, click Geometry. Click Tube. In the Top viewport, create a short, thin tube.

Under the Modifier panel, click Edit Mesh. In the Front viewport, select the top set of vertices on the tube.

Choose Select and Rotate. 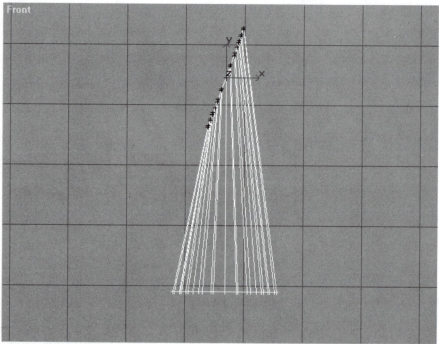 In the Front viewport, rotate the vertices by about 70 degrees. Watch the status line to see how far the vertices are rotated.

*Figure 3-39. Rotated vertices*

Right-click on the Top viewport to activate it.

Choose Select and Non-uniform Scale. 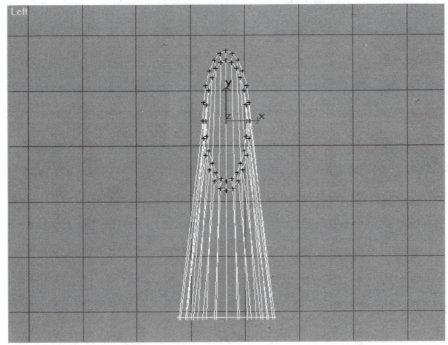 Turn on Restrict to

Y. In the Top viewport, scale the selected vertices to about 50%.

Right-click in the Left viewport to activate it. Make sure Restrict to Y is on. Scale the selected vertices to about 150%.

*Figure 3-40. Scaled vertices*

Select the bottom vertices in the Front or Left viewports.

Choose Select and Uniform Scale. 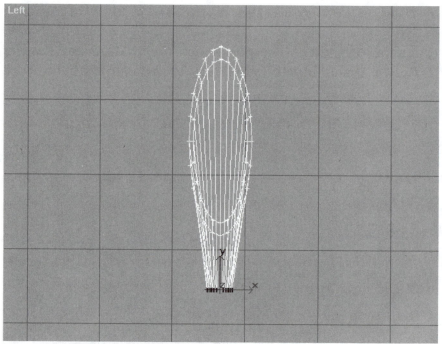 In the Top viewport, scale the vertices to about 20%.

Also move the vertices up as shown in Figure 3-41.

Figure 3-41. Scaled and moved vertices

The ear is now complete. Turn off Sub-Object.

Rotate and move the ear to position it on the kangaroo head, as shown below. Scale the ear if necessary to suit the size of the kangaroo. The base of the ear should be firmly inside the kangaroo head.

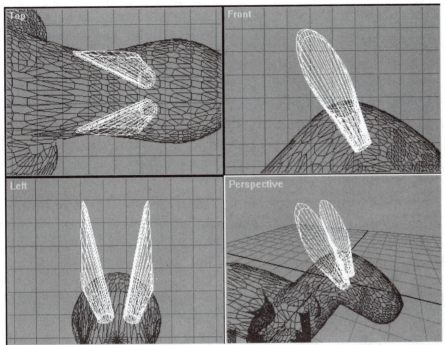

Figure 3-42. Ear position

With the ear selected, click on Mirror Selected Objects. ▶◀ Turn on Copy under the Clone Selection, and click OK to create the mirrored copy. Position the mirrored ear on the other side of the head.

Under the Display panel, choose Unfreeze All. Select the body. Under the Modify panel, choose Attach Multiple from the Edit Object rollout. Select the two ear tubes.

Save the file as KANGB06.MAX.

Select the kangaroo body. Under the Display panel, choose Freeze Selected.

### Step 24. Eyes and Eyelids

To create the eyes, create a small sphere. Name the sphere EyeR. Position the sphere against the head as shown in Figure 6-43.

Choose Select and Uniform Scale. 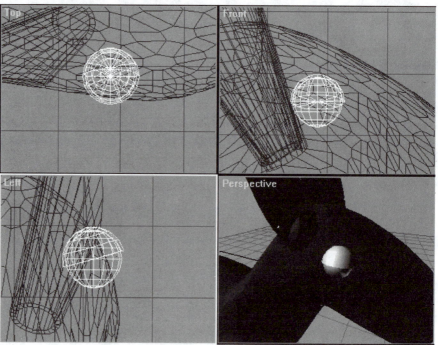 Hold down the <Shift> key and scale the eye up to about 110%. Name the new object EyelidR.

Under the Modify panel, change the Hemisphere value to 0.5. This turns the sphere into a hemisphere.

Rotate the hemisphere so the eye and eyelid look like Figure 3-43.

Figure 3-43. Eye and eyelid position

Select the two spheres. Activate the Top viewport.

Choose Select and Move. Select both eyes and both eyelids. Mirror-copy the eyes to the other side of the head. Name the new objects EyeL and EyelidL.

Unfreeze the kangaroo body. Select all eyes and eyelids. Choose

Select and Link. Click and drag from the eyes and eyelids to the body to link them to the body.

Save the file as KANGB07.MAX.

## Step 25. Materials

A bitmap has been supplied for use on the kangaroo body. This bitmap is an altered version of the bitmap PELT.JPG that comes with MAX. The file, called NEWPELT.JPG, can be found in the \SANFORD\MAPS directory on the CDROM.

To make the pelt material, use NEWPELT.JPG as a Diffuse map and apply the material to the kangaroo body. You will need to apply mapping coordinates to the body in order to use this material. Use planar UVW mapping applied to the side of the body, fitting the gizmo to the kangaroo's body.

The same material can be applied to the eyelids using the same mapping coordinates. Make the eyes a plain black or brown color.

*In the model shown in Figure 3-44, the hands were rotated inward to simulate a kangaroo's natural pose.*

Figure 3-44. Kangaroo with materials

Save the file as KANGB08.MAX.

The kangaroo body is now complete.

Go on to the next tutorial to find out how to make the kangaroo jump.

## Tutorial 8
## Jumping with Character Studio

Making a kangaroo jump is no easy task with standard MAX tools. When you need to make a character walk, jump, dance or otherwise act naturally, it's essential that you use a MAX plug-in called Character Studio.

A Character Studio working demo is included with the 3D Studio MAX 1.1 Upgrade CDROM. Although the demo version of Character Studio will allow you to load and view biped files, it won't let you create or save bipeds. You cannot do this tutorial unless you have a full working version of Character Studio. However, you can load and view the sample files on the CDROM that comes with this book.

*You must have a working version of Character Studio to do this tutorial.*

Character Studio consists of two modules: Physique for creating characters, and Biped (pronounced *by'ped*) for animation. Both modules are used in this tutorial.

Character Studio works with ordinary MAX models. Once you create a character model in MAX, you can use Biped to animate the model with any of the pre-programmed motions that come with Character Studio. You can also create a custom animation through small icons called *footsteps*. After the footsteps are placed, Character Studio automatically makes the character move by placing its feet in the footsteps, much like learning to dance.

The kangaroo jump requires custom animation. In this tutorial, you'll place footsteps for the kangaroo and make it jump across the screen. Animating with footsteps might seem awkward at first, but the confusion turns to pleasure after you see how rapidly a character can be animated. Once the basic jumping animation is finished you can animate the head, arms, legs and tail to add realism to the kangaroo's movements.

To help you start animating with Character Studio as quickly as possible, two files have been supplied on the CDROM that comes with this book. All can be found in the \SANFORD\MESHES directory.

**KANGBIPD.MAX**      **Kangaroo biped**
**KANGJUMP.MAX**      **Complete animated scene**

KANGJUMP.MAX has the kangaroo jumping along in the Australian outback, finding a tree and eating some leaves.

There are some significant differences between this kangaroo and the kangaroo you built in the modeling tutorial. Although a basic animal head can be created when lofting the body as one piece, a more detailed head is required for close-ups. On my kangaroo, I created and attached a new head that's closer to what a real kangaroo head looks like. The new head was lofted with fit deformation in the same manner as the body, but with a number of cross sections. I then pushed and pulled vertices to define the hollows of the kangaroo's face.

To attach the new head to the kangaroo mesh, I had to first remove the old one. I used Edit mesh to select the vertices of the original head and delete them from the rest of the kangaroo body. Then I positioned the new head and joined it to the neck with a boolean union. After the head was finished I created a separate jaw by flattening a sphere, then added nostrils, eyes, and eyelids. I also bent the kangaroo forward to suggest a kangaroo's pose just before jumping. You will find these extra parts and the new pose in KANGJUMP.MAX.

In the finished scene, the kangaroo mesh was attached to the biped with Physique. However, the jaw, nostrils, ears, hands and feet were linked directly to the biped, not attached to the body mesh. I have found that by not using Physique on the small parts, there are fewer problems with vertex assignments.

You can make the entire model in one piece, but because boolean operations in MAX are sometimes difficult on complex objects, I use booleans only when it's necessary to have a seamless mesh. When watching the animation, you probably won't notice the seams where the hands meet the arms because they're always in motion.

There are many unique concepts in Character Studio. For best results, please take the time to go through Tutorials 1, 2, and 3 in the Character Studio manual before doing this tutorial.

### Step 1. Load biped

For this tutorial, a biped has already been reshaped to fit the kangaroo mesh.

Load the file BIP01D.MAX from the \SANFORD\MESHES directory on the CDROM. This file contains my more detailed kangaroo model, and a biped to fit it.

Figure 3-45. Kangaroo and biped

Click Select by Name ![icon] to see the list of objects in the scene.

In addition to the kangaroo body mesh, this file also contains the kangaroo ears, eyes, hands and feet as separate objects. The names of these additional objects have brackets around them so you can find them easily at the top of the list.

The name of each object in the biped begins with the prefix Bip01. The kangaroo biped includes three neck links, five tail links and several links for the toes and fingers. Zoom in on the hands and feet to see the relationship between the biped and the mesh.

If you were creating a biped from scratch, you would move, rotate and scale the biped parts to match the kangaroo's body. This step has been done for you already.

### Step 2. Link body to biped with Physique

Before you can use the biped to animate the kangaroo, the biped must be linked to the kangaroo mesh. This tells Character Studio that you want the biped links to correspond to nearby vertices on the kangaroo, and to deform the kangaroo mesh accordingly in the final animation.

Only the kangaroo mesh is to be connected to the biped with Physique. To ensure they aren't accidentally connected to the biped, the additional body parts will be hidden.

The mesh will be linked to the object Bip01 Pelvis. To make the linking step easier, hide all objects except KangMesh and Bip01 Pelvis.

Select the kangaroo mesh. Under the Modify panel, click More and choose Physique from the list. Wait a few moments while Physique opens.

*A Biped skeleton is made up of links and nodes. A link behaves like a bone, and a node is a joint between any two links that make up a Biped skeleton. When you click on a node to rotate the lower arm, the elbow node (joint) is activated and the arm is rotated around the node's axis.*

*The kangaroo mesh must be selected or Physique will not open.*

133

Zoom in on Bip01 Pelvis in the Top viewport. Under the Physique rollout, click the Attach To Node button. Move the cursor over Bip01 Pelvis in the Top viewport. The cursor changes to look like the Attach to Node button to indicate a legal link can be made. Click to set the link.

Wait a few moments while Physique calculates all the links from the mesh to the biped skeleton. When linking is complete, a set of orange lines will appear through the model.

Click Zoom Extents All.

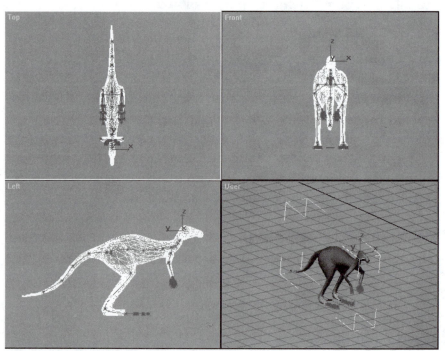

*Figure 3-46. Mesh linked to biped with Physique*

## Step 3. Link kangaroo parts

Now you are ready to attach the the hands, feet, ears and eyes directly to the biped with MAX's standard linking command.

Hide all objects currently on the screen. Unhide the eyes, ears, hands and feet. Also unhide the objects Bip01 R Hand, Bip01 L Hand, Bip01 L Foot, Bip01 R Foot and Bip01 Head.

Zoom in on the hands in the Front viewport. Select the object LHand. Click Select and Link. Click and drag LHand and move the cursor over the corresponding biped hand link Bip01 L Hand. The top square on the cursor will turn green to indicate a legal link.

*In the Front viewport, the kangaroo's left hand appears on the right, and the right hand appears on the left.*

Release the mouse button. The hand is now linked to the biped.

Zoom in on objects as necessary to link the remaining mesh objects to the biped objects as follows.

| Mesh object | Link to biped object |
| --- | --- |
| RHand | Bip01 R Hand |
| LFoot | Bip01 L Foot |
| RFoot | Bip01 R Foot |
| Eyes | Bip01 Head |
| Ears | Bip01 Head |

### Step 4. Place initial footsteps

Next you'll place footsteps in the scene to make the kangaroo jump. Making the kangaroo perform a basic jump can be accomplished with relatively few steps. The built-in Inverse Kinematics of the biped will automatically maintain its balance as it moves from one step to the next.

Footstep animation can be created either before or after the mesh skin is attached to the biped. I prefer to do it afterward.

The first motion of the kangaroo will be a series of four jumps, each 20 frames long. As you create new footsteps, the length of the total animation will be increased automatically by Character Studio.

Hide all objects. Unhide only the biped objects for this step, all objects beginning with Bip01.

Click the Top viewport. Zoom out to give yourself some room to work. Pan the viewport upward with the biped still visible, but with plenty of space in front of the biped in the direction of travel.

Make a second Top viewport by converting the Front viewport to a Top viewport. Do this by clicking on the Front viewport and pressing the <T> key. In this second Top viewport, zoom in close to the biped's feet so that you can see them easily.

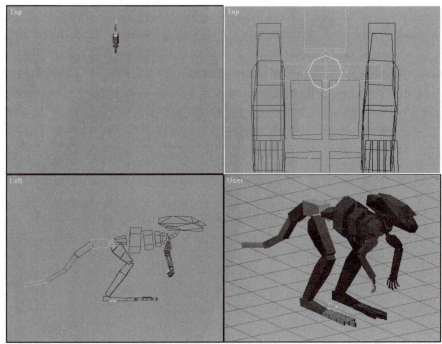

*Figure 3-47. Viewport setup for footstep creation*

Select Bip01. Go to the Motion panel. Under the Track Selection

rollout, click the Footstep Track button. ![footstep icon] This will enable footstep creation.

**TIP** *Do not turn on the Animate Button when animating with footsteps.*

Under Footstep Creation, click the Jump button, 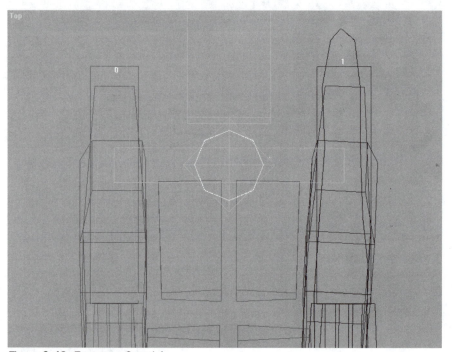 then click the Create Footstep (at current frame) button . The mouse pointer turns into a footstep cursor. In the second Top viewport, locate the biped's right foot, which is the leftmost foot in the viewport. Click on the heel of the leftmost foot. This will create footstep 0. Then click on the heel of the biped's right foot. This action creates footstep 1.

Figure 3-48. Footsteps 0 and 1

Now there are two elongated footprint shapes side by side, labeled with the numbers 0 and 1. The two footstep objects define the starting position of the biped's motion.

Click Select and Move, then click and drag the footsteps to align them to the feet of the biped.

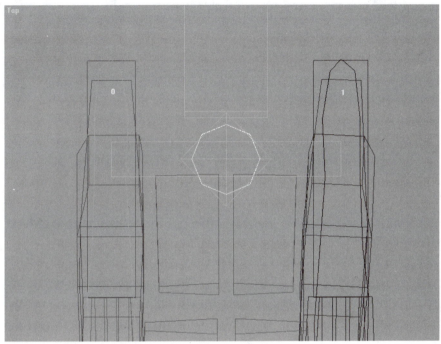

*Figure 3-49. Aligned footsteps*

**TIP** *When placing the new footsteps, be sure to maintain approximately the same spread and alignment as the first two footsteps so that the feet will stay together during the jump.*

## Step 5. Place all footsteps

Note that by default, the 2 Feet Down value is 9 and the Airborne value is 11. This means that each new pair of footsteps in Jump mode will contact the ground for 9 frames, and be airborne for 11 frames.

Activate the first Top viewport and click Min/Max Toggle. Under the Footstep Creation rollout, click Create Footsteps (append). Move the cursor to a location lower on the screen directly in front of the biped, just below the biped's head. Try to place the cursor directly below the leftmost foot.

Click to place Footstep 2. Move the cursor to the right a short distance and click to place Footstep 3.

Place three more sets of footsteps. Place the footsteps apart by approximately 1-1/2 times the length of the kangaroo's body. There should be a total of 10 footsteps numbered 0 through 9.

*Figure 3-50. Footsteps for hop*

You can use Select and Move to move the footsteps into the correct positions after you've created them.

Click Min/Max Toggle to return to the four-viewport display.

## Step 6. Create keys

Once you've placed all the footsteps, one additional command will cause biped to create motion keys for the biped automatically.

Under the Footstep Operations rollout, click on Create Keys for Inactive Footsteps. This creates an animation track based on your footsteps. The blue and green footprints will change color, indicating that they are now active and Keyframes have been created by Character Studio.

If you wish to see the new Footstep Track Keys, can now open the Track View and expand the Object Hierarchy to see the Bip01 Footstep Transform Keys. They are represented by numbered green and yellow squares in the Track View Window. Close the Track View after you have inspected the Footstep Track.

## Step 7. View and adjust animation

*If your biped crouches too much during the first hop, this means the first two pairs of footsteps are too far apart. Use Select and Move to move footsteps 2 and 3 closer to footsteps 0 and 1.*

In order to see the biped move, Character Studio provides a special Quick Preview button in the Motion Panel biped menu. Click on the Left viewport and click Zoom Extents so you can see all the numbered footsteps.

In the Motion Panel biped rollout, under General, click Biped Playback. A stick figure representing the biped skeleton will move in real time in the Left Viewport, showing the biped jumping. Watch the animation in the Left viewport. To stop the playback, click the blue biped Playback Arrow again. Click MAX's standard Go to Start button to return to frame 0.

*Figure 3-51. Biped jumping*

The biped jumps, but not very high. This is a very sluggish kangaroo! To make the kangaroo jump higher, change the GravAccel spinner. Increase the number by 50%, then play the animation again. The kangaroo jumps with more height and grace.

Keep trying different values until the jump height is about half of the kangaroo's height. To save the motion information, click in the General rollout in the Motion Panel on the Save File button. Save the motion file as KANG.BIP.

Before going to the next step, hide the footsteps in the viewport by clicking on Show/Hide Footsteps.

### Step 8. Render preview

Now you're ready to see how the skinned kangaroo moves.

Hide the biped, and unhide all kangaroo body parts.

Activate the Left viewport. Under the Rendering menu, choose Make Preview. Click Active Time Segment, then click on Create. Watch the preview render, and then play it back with the Media Player.

Check for problems such as the feet moving into the wrong position or crossing over each other. Because the footsteps are actually objects, you can move them like any other objects and correct these problems.

To move the footsteps, return to the Motion panel. Activate the Footstep Track. Use Select and Move to reposition them as necessary. Play the animation using biped Playback. Be sure to close the Motion Panel and save the MAX scene file after you modify the footstep placement.

### Step 9. Vertex link adjustment

When you inspect the model at different places in the animation, you might find some areas of the kangaroo where vertices are sticking out and forming a lump, such as around the area where the arms meet the body. This is caused by Physique having assigned these vertices to the wrong link. In this step you will learn how to correct these problems.

Each time I have done a new version of the kangaroo, the Physique vertex assignment has shown a different set of vertices linked incorrectly.

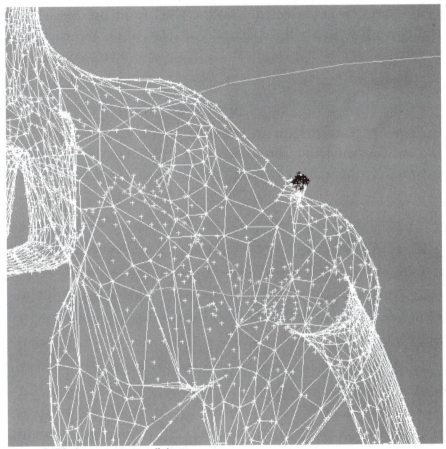

*Figure 3-52. Improper vertex linkage*

Figure 3-52 shows what can happen when the vertices that would normally be connected to the spine are instead connected to another link somewhere else in the biped.

To correct the vertex linkage problem shown in Figure 3-52, you would select the kangaroo mesh, then click the Modify panel. This will automatically turn on Physique. Zoom in on the problem area in any viewport and maximize the viewport. You may have to use Arc Rotate Selected to get an unobstructed view of the problem area.

Click Sub-Object. Select the Vertex level from the Sub-Object pulldown list. You will experience a delay while the computer calculates the vertex positions. The Physique Vertex Assignment rollout appears.

Click the Select button under Vertex Operations. The button will turn green. Now click on the problem vertex, the one that's sticking out past the kangaroo's side. After a moment, the vertex will turn red. If there are a group of vertices giving you trouble, you can hold down <Ctrl> and continue clicking until you have selected them all.

A vertex that turns red when selected indicates that the vertex is a deformable vertex, meaning it is part of the moveable skin. If the vertex turns blue, it means that Physique has accidentally assigned it to the root object Bip01.

If the vertices are red, click Assign to Link, then click on the nearest yellow link that makes up the spine. The vertices will pop back into the mesh and will no longer protrude.

For detailed information on changing vertex assignment that turn blue when selected, refer to the Character Studio manual.

Save your .MAX file after correcting the vertex assignments.

# Things to Try

There are many improvements that can be made to the kangaroo animation. Here are a few.

## Realistic Motion

What makes good character animation so special is the animator's attention to nuance and subtlety. The motion supplied by Character Studio's default settings is somewhat mechanical, but all the tools are there for you to give the character more personality and life.

The motion you set up for the kangaroo in the last tutorial is okay for starters, but the animation would benefit from some subtle changes. However, a detailed tutorial on how to make the kangaroo move realistically would take up an entire book in itself.

When animating a biped model, you'll do well to look at motion in life. For the kangaroo animation, for example, I watched nature videos of kangaroo herds. If you're animating a person, move around yourself and watch the motion, or get a friend to help you.

The small motions you pick up — a hand movement here, a turn of the head there — will breathe life into your animation and make it really special.

## Movement Along Terrain

A biped can be made to follow a contoured surface by simply moving the footsteps to match the contours. The biped's toes will not pass below the footstep objects.

You can create a contoured ground surface by applying a Noise modifier to a large box. You can also create a terrain model with Tutorial 22 in this book.

## Rendering

To make a final rendering, use a background image that suggests the Australian outback such as a picture of desert hills and sky.

Place subdued omni lights in the scene. Add one strong Directional Spotlight overhead to cast the kangaroo's shadow on the ground.

This project is ongoing and will eventually contain a herd of kangaroos eating and hopping around to the music of an Australian Diggery Do.

## Sample Animation

KANGJUMP.MAX file contains several additional elements. In the file there is a tree so the kangaroo has some lunch to eat. The kangaroo jumps along, sees the tree, then stops and eats a handful of leaves. The file has been rendered as KANGJUMP.AVI which can be found in the \SANFORD\AVI direcotry on the CDROM.

In this project, I have separated the action into three parts. Camera01 shows the kangaroo jumping along, and the camera pans right to follow him.(frames 1 to 165) Camera02 shows the kangaroo approaching the tree and inspecting it to decide which leaves to eat. (frames 167 to 240) Camera03 shows a close-up of the kangaroo from a new angle standing beside the tree and eating a few leaves (frames 282 to 332). In this third section the kangaroo's hands and jaw are animated so that he bites the branch held in his hands, and eats the leaves.

All maps used in these sequences can be found in the \SANFORD\MAPS directory on the CDROM.

# chapter

## High Heeled Shoe

To make a ladies' high heeled shoe in MAX, you might think you should work like a shoemaker, cutting and stitching the individual pieces together. But unless you've spent a few years as a cobbler's apprentice, it's difficult to visualize the pieces necessary to make such a shoe.

In attempting to make a high heeled shoe I experimented with several methods in MAX, and finally hit on a way that worked. I used fit deformation to create the general shoe shape, then applied a boolean subtraction to make the empty space for the foot to slide in. The heel was created separately. This method is detailed in the tutorial that follows.

Fit deformation is a lofting technique. Instead of making numerous cross section shapes, fit deformation makes life easier by using an object's profiles. One general shape or shapes are placed along a path, then the top and side view of the object are used to generate more detailed cross sections.

More information on fit deformation is included at the beginning of Chapter 3.

*Figure 4-1. The finished shoe*

To prepare for this tutorial, I took a shoe from my closet. When I look at the shoe from the top, the profile looks like Figure 4-2.

*Figure 4-2. Top profile of shoe*

When I look at the shoe from the side, the profile of the shoe without the heel looks like Figure 4-3.

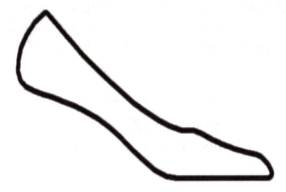

Figure 4-3. Side profile of shoe

These profiles will be used in the tutorial that follows.

### Tutorial 9
# Shoe

To make the high heeled shoe, you'll start by making the body of the shoe itself. Fit deformation and booleans will be used for this purpose. Later on you'll make the heel and assign materials.

### Step 1. Create top shape

First you'll create a shape for the top profile of the shoe. The shoe will face sideways, with the heel at the left and the toe at the right.

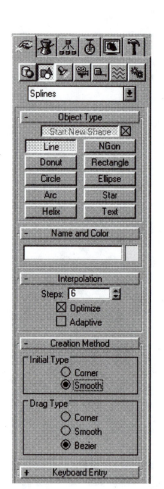

Click on the Top viewport, then click Min/Max Toggle to make the viewport fill the screen.

Under the Create panel, click on Shapes. Click the Line button. Under the Creation Method rollout, under Initial Type, turn on Smooth. In the Top viewport, create a shape with five vertices similar to Figure 4-4.

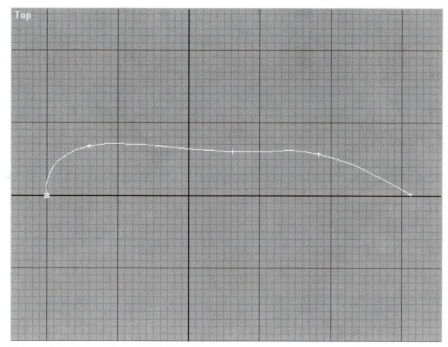

Figure 4-4. Smooth line with five vertices

Try to make the first and last vertex line up along the horizontal. The easiest way to do this is to line each vertex up with one of the grid lines. The vertices don't have to line up precisely, but they should look to the naked eye as though they're sitting on the same gridline.

If your shape doesn't look like the shape above, you can edit it. Under the Modify panel, click on the Edit Spline button. The Sub-Object level is already on, and the Vertex level is selected.

Click on Select and Move, and click on a vertex to select it. Click and drag the vertex to move it. Move the vertices until the shape looks like Figure 4-4.

When you've finished, be sure to click on Sub-Object again to deselect it.

Next you'll mirror the shape to make the other half of the Top profile. Under the Modify panel, click on the Edit Spline button. From the Sub-Object pulldown list, choose Spline.

A new set of rollouts appears. Pan down to see all the options under the Edit Spline rollout. The Mirror option will be used to mirror and copy the spline.

Next to the Mirror button, turn on the Copy checkbox. Under the Mirror button are three buttons for setting the type of mir-

ror. Click on the Mirror Vertically button to turn it on.

In the Top viewport, select the spline, then click the Mirror button. A mirror copy of the spline appears on the screen.

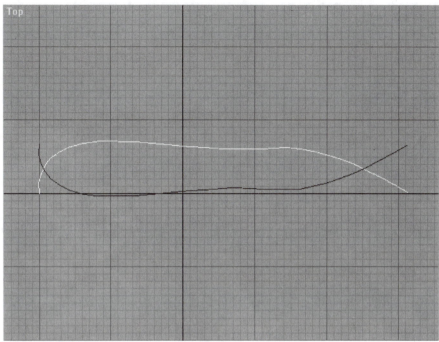

*Figure 4-5. Spline copied and mirrored*

Choose Select and Move. 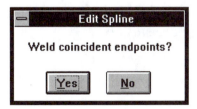 Turn on Restrict to Y ![Y] and move the spline downward until the end vertices touch. A dialog box appears.

---

**Edit Spline**

**Weld coincident endpoints?**

Yes    No

---

Click on Yes. The two spline halves are welded together.

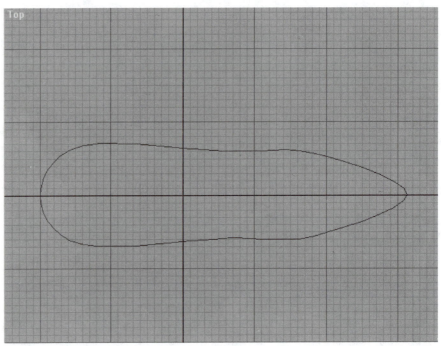

*Figure 4-6. Welded spline*

Right now the shoe is symmetrical, which would make it uncomfortable to wear. A few minor modifications will make the shape into a left shoe.

On the Sub-Object pulldown list, choose Vertex. Note that Select and Move  and Restrict to Y  are still on.

In the Top viewport, move vertices along the Y axis only until the shape looks like Figure 4-7. Name the spline ShoeTop.

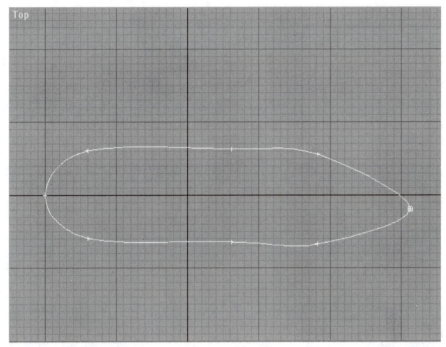

Figure 4-7. Final top profile

The top profile is now complete.

### Step 2. Create side profile

The side profile will be created just underneath the top profile.

Click Pan 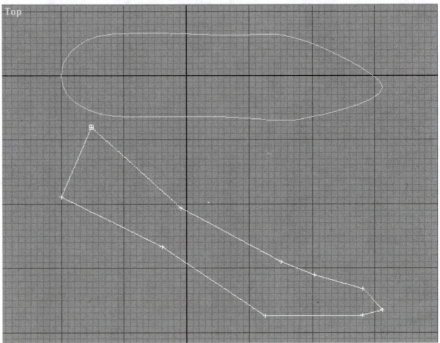 and slide the viewport display upward to make room for the new shape.

Under the Create panel, click Shapes, then the Line button. Under the Creation Method rollout, under Initial Type, turn on Corner. Create a spline similar to the one below.

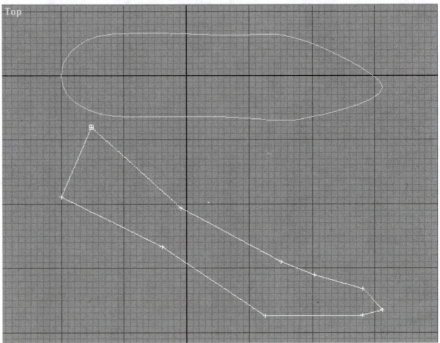

*Figure 4-8. Spline with corner vertices*

Now you must edit each vertex to make a smooth shoe profile.

Under the Modify panel, choose Edit Spline and turn on Sub-Object. Make sure Vertex is selected from the Sub-Object pulldown

list, and click Select and Move.  Turn on Restrict to XY
Plane.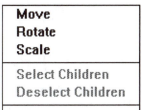

Right-click on the upper left vertex of the spline. Choose Bezier Corner from the pulldown list.

Two handles appear around the vertex. Move the handles so they look like Figure 4-9.

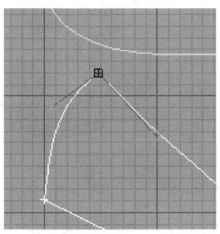

*Figure 4-9. Bezier Corner handles positions*

Next, right-click on the vertex at the heel. Choose Bezier. Two handles appear around the vertex. Move the handles so the curve looks like Figure 4-10.

***To move each handle separately, hold down the <Shift> key and click and drag on one of the handles.***

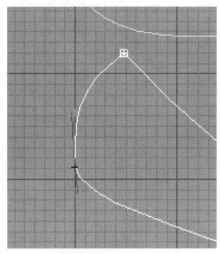

Figure 4-10. Bezier handles positions

Change all other vertices to Bezier. Adjust the handles on each vertex until the spline looks like Figure 4-11. Name the spline ShoeSide.

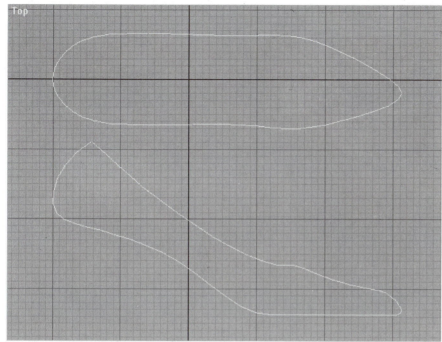

*Figure 4-11. Side view spline under top view spline*

Both the top and side profiles were created with a vertex at the very left of the shape and a vertex at the very right. Fit deformation works best when profiles are made in this way.

The side view spline for the shoe is now complete.

### Step 3. Create path

A path is needed for the fit deformation. The path must be the same length as the two profiles.

Under the Create panel, click Shapes. Choose Line. Create a straight line under the side view shape. Make the line the same length as the top and side profile splines. Leave some room between the side view spline and the path.

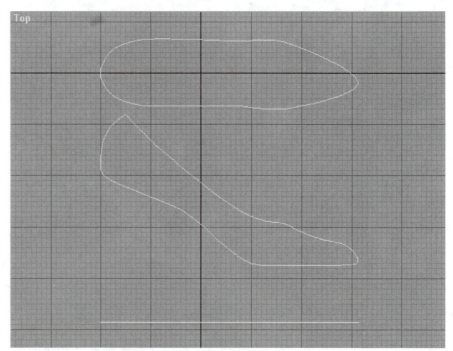

*Figure 4-12. Path spline under top and side view*

Change the name of the path to OuterPath. This path will be used to create the outer shoe mesh for the boolean operation.

The path needs some detail that will be useful later on. Under the Modify panel, choose Edit Spline. Click on Sub-Object to edit at the Vertex level.

Click on Refine. Place three more points along the path as shown in Figure 4-13. After placing each vertex, right-click on the vertex and choose Corner as the vertex type.

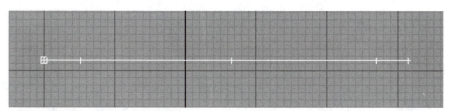

Figure 4-13. Extra vertices on path

Turn off the Sub-Object button.

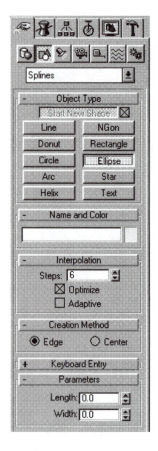

## Step 4. Create shape

A general shape is needed to pass along the path. A modified ellipse will do for this shape.

Under the Create panel, click Shapes. Choose Ellipse. Click and drag in the Top viewport to create an ellipse, as shown in Figure 4-14.

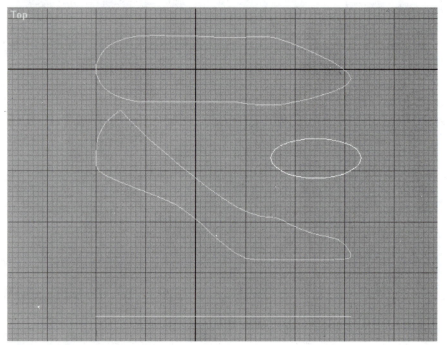

*Figure 4-14. Ellipse*

Now you'll modify the ellipse to give it a rectangular shape. Under the Modify panel, choose Edit Spline. Click on Sub-Object and make sure Vertex is selected as the Sub-Object level.

Click on the top vertex of the ellipse, then hold down the <Ctrl> key and click on the bottom vertex. Both vertices are now selected, and their Bezier handles appear.

# Master Pieces

# Sanford Kennedy
## Tutorials 7 & 8

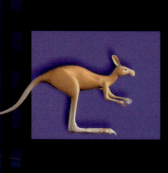

# Ted Boardman
## Tutorial 6

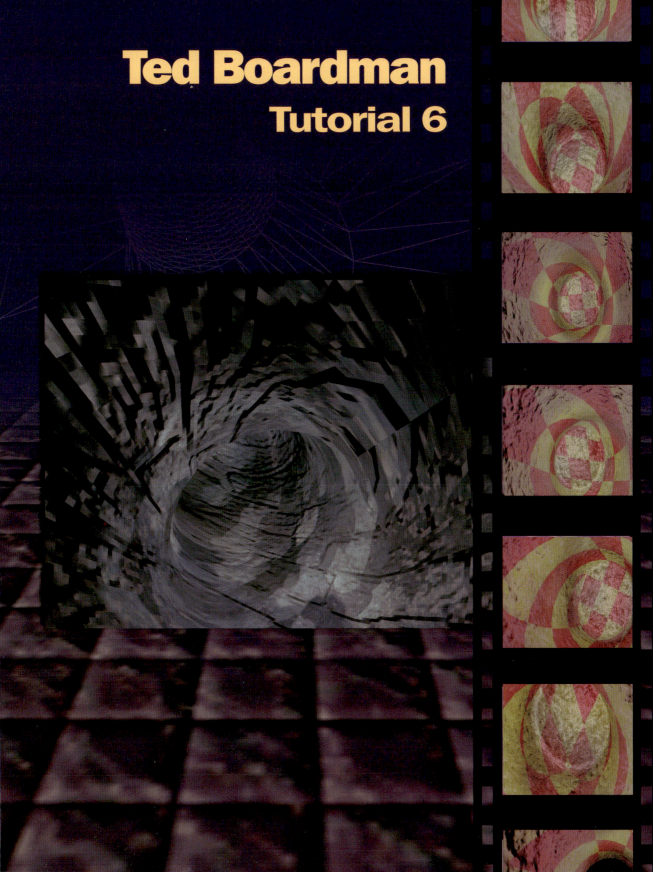

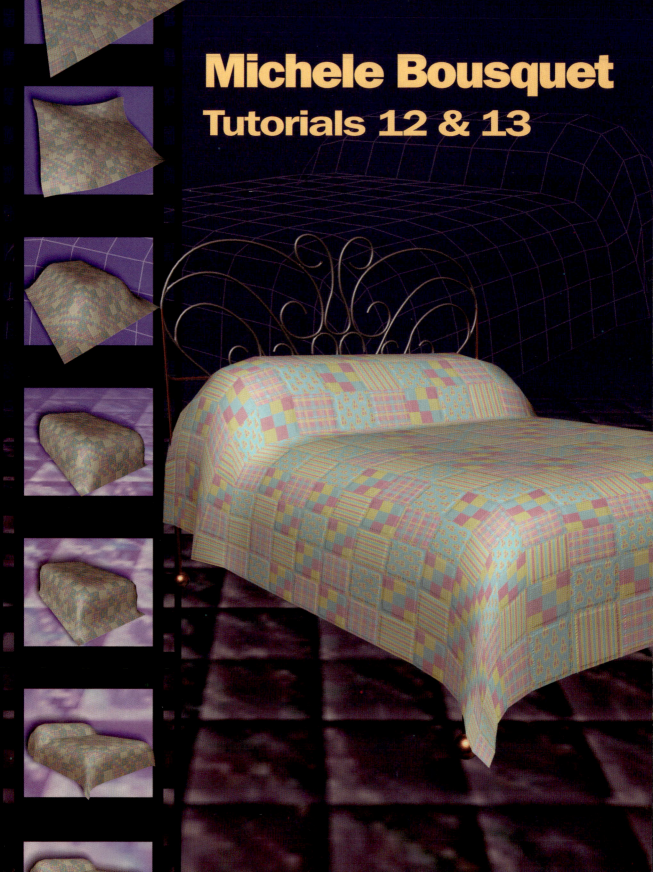

# Michele Bousquet
## Tutorials 12 & 13

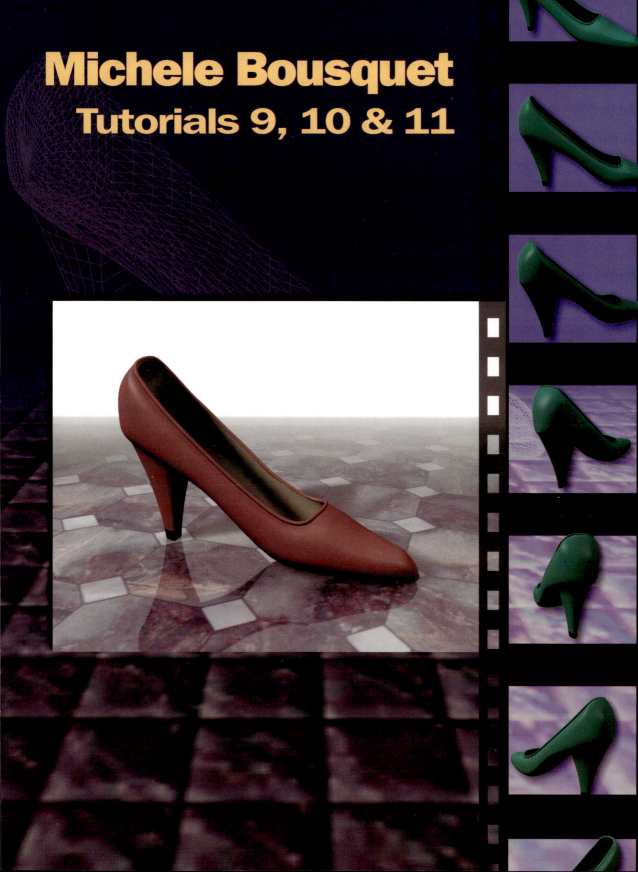

# Kyle McKisic
## Tutorial 15

# Kyle McKisic
## Tutorials 16 to 20

# Frank Delise
## Tutorial 14

At the bottom of the Edit Vertex rollout, turn on the Lock Handles checkbox and the All checkbox.

Turn on Restrict to X. ![X] Click and drag any one of the Bezier handles displayed on the ellipse. All the Bezier handles move at the same time. Move the handles until the ellipse looks like Figure 4-15.

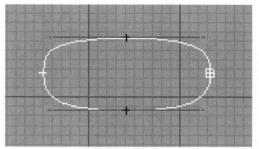

Figure 4-15. Modified ellipse

> **TIP** Be sure not to pull the handles past the left and right vertices. If you pull the handles too far, the shape will indent at the sides like an hourglass, causing the final shoe model to look very strange.

Turn off the Sub-Object button.

## Step 5. Loft outer shoe

You now have all the shapes you need to make the outer shoe mesh for the boolean operation. The mesh will be created with lofting and fit deformation.

Select OuterPath. Under the Create menu, click the Geometry button. Select Loft Object from the pulldown list.

Under the Object Type rollout, click the Loft button. A new set of rollouts appears. Under the Creation Method rollout, click Get Shape. Click on the ellipse. A copy of the ellipse moves to the left end of the path.

Under the Skin Parameters rollout, click on Skin. The lofted object appears in the viewport. Click on Zoom Extents 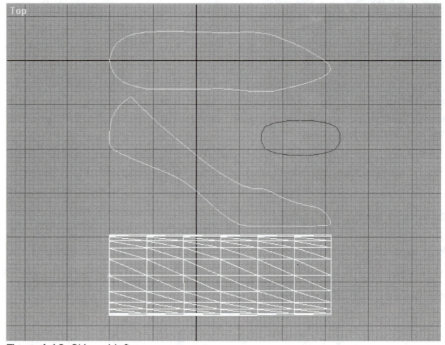 if necessary to see the lofted object.

Figure 4-16. Skinned loft

Click Min/Max Toggle ![Min/Max Toggle icon] to go back to the four viewport display. Click Zoom Extents All ![Zoom Extents All icon] to see the lofted object in all four viewports.

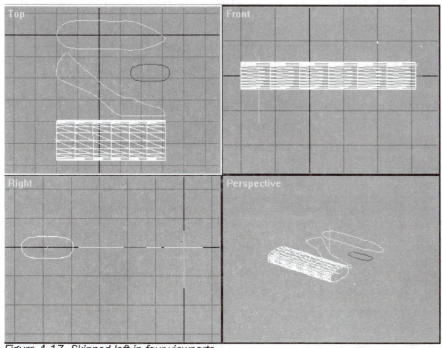

*Figure 4-17. Skinned loft in four viewports*

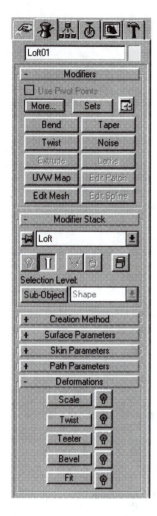

## Step 6. Fit deformation

The loft doesn't look much like a shoe, does it? Well, you were promised a shoe, and you shall have a shoe.

With the loft object still selected, go to the Modify panel. A new set of rollouts appear for modifying a loft object. Many of the rollouts are the same as those that appeared when you created the loft object, but there is one new one. The bottom rollout is called Deformations.

Expand the Deformations rollout. Click on the Fit button. The Fit Deformation window appears. This is the window to which you will bring in the top and side profiles to deform the loft object.

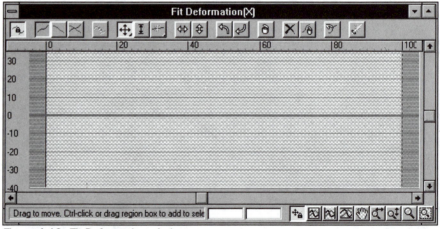

*Figure 4-18. Fit Deformation window*

Turn off the Make Symmetrical button. This will allow you to bring in two different shapes for the top and side profiles.

By default, when the Fit Deformation window appears it's ready for the X profile. Click on Get Shape, 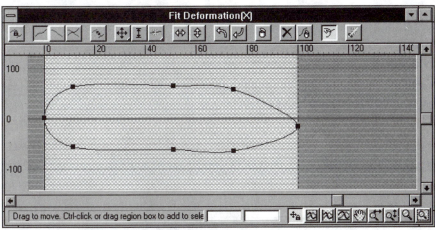 and click on the top profile spline, ShoeTop, in any viewport. The profile appears in the window. To see the entire profile, click the Zoom Extents

button ▨ at the bottom of the Fit Deformation window.

*Figure 4-19. X profile in Fit Deformation window*

Move the Fit Deformation window aside to see the Top viewport. The loft object has changed to look like the X profile, when viewed from the Top viewport.

Move the Fit Deformation window so you can see all the buttons

on the window. Click on Display Y Axis. ◿ The X profile disappears. The Fit Deformation window is now prepared to accept the Y shape.

The Get Shape button is still pressed, so there's no need to press it again. Click on the side profile shape, ShoeSide, in any viewport. The Shape appears in the window. Click Zoom Ex-

tents ▨ to see the entire shape.

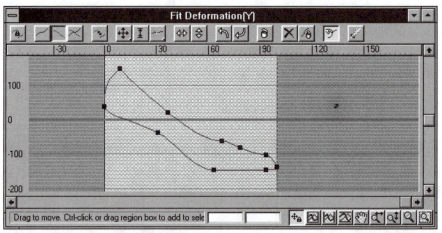

*Figure 4-20. Y profile in Fit Deformation window*

Close the Fit Deformation window. In all four viewports, you can see that the lofted shape has been altered to conform to the top and side profile shapes.

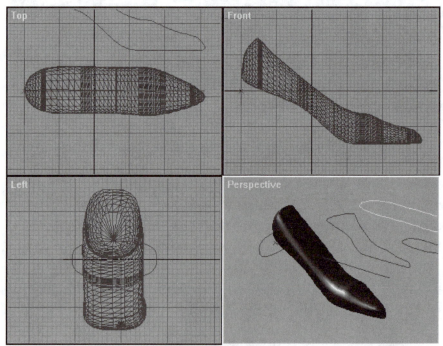

*Figure 4-21. Outer shoe created with fit deformation*

Change the name of the object to OuterShoe. This object will be used as the outer shoe for the boolean operation.

## Step 7. Redistributing detail

Currently, the path steps coincide with the vertices on the profiles. This gives you a lot of unnecessary detail at some parts of the shoe.

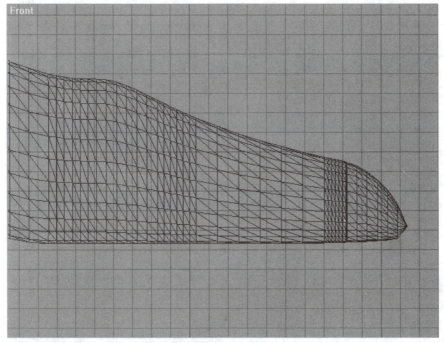

*Figure 4-22. Poorly distributed shoe detail*

The shoe needs a lot of detail at the heel and toe but only moderate detail along the middle.

Putting in detail only where needed will not only make redraw and rendering times faster, but will cause the boolean operation to perform faster, too.

To change the distribution of faces along the shoe, select the shoe and go to the Modify panel. Expand the Skin Parameters rollout. Watch the Front viewport and turn off the Adaptive Path Steps checkbox. The path steps now coincide with the vertices you placed along the path when you created it, rather than by the vertices on the profile shapes.

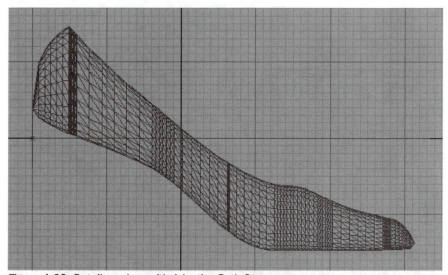

*Figure 4-23. Detail on shoe with Adaptive Path Steps on*

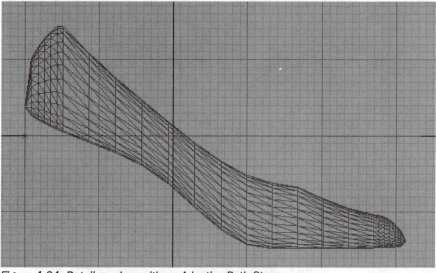

*Figure 4-24. Detail on shoe with no Adaptive Path Steps*

There aren't quite enough path steps. Change Path Steps to 8. The shoe now has sufficient detail along the path.

In the Left viewport, you can see the upper part of the back of the shoe. The level of detail that appears in the viewport suggests that there are not enough shape steps around the object to make a smooth shoe. Change Shape Steps to 8 as well.

The outer shoe now has the minimum detail necessary to be smooth. Click on the Perspective viewport and select OuterShoe.

Click Zoom Extents Selected.  Render the Perspective view.

Figure 4-25. Rendering of outer shoe

### Step 8. Top profile for cutout

Next you'll create another object to cut out the inside of the shoe. You can create copies of the existing shapes to make your work easier. In MAX, any object or spline can be automatically copied by holding the <Shift> key while transforming the object.

Click in the Top viewport to activate it. Click on Min/Max Toggle to enlarge the Top viewport.

Select the top profile, ShoeTop. Choose Select and Non-uniform Scale. Click on Restrict to X.

*While scaling the spline, watch the status line at the bottom of screen to see the scale percentage.*

Hold down the <Shift> key and scale ShoeTop to 95%. When you release the button, a dialog box appears.

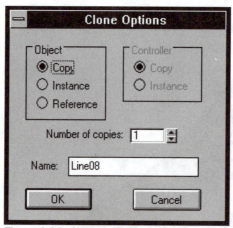

Figure 4-26. Clone Options dialog box

Enter a new name for the new spline, ShoeTopCutout.

Turn on Restrict to Y. Scale ShoeTopCutout on the Y axis to 80%.

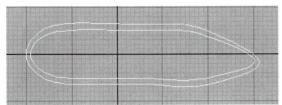

*Figure 4-27. Top profile for cutout*

## Step 9. Side profile for cutout

The side profile for the cutout must be made with care. This shape must jut out over the top of the shoe to cut out the opening for the foot to slip into.

Select the side profile ShoeSide. Choose Select and Move  and turn on Restrict to Y. Hold down the <Shift> key and move the spline up by a few units so the sole sits just inside the shoe, as shown in Figure 4-28. Name the new spline ShoeSideCutout.

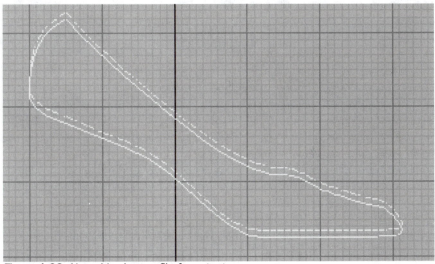

*Figure 4-28. New side view profile for cutout*

Choose Select and Non-uniform Scale 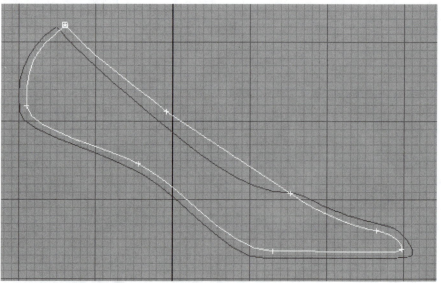 and turn on Restrict to X. Scale the new spline along the X axis to 95%.

Next you must adjust some of the vertices. Under the Modify panel, choose Edit Spline. Click on Sub-Object and make sure Vertex is selected. Click on Select and Move and turn on Restrict to XY. Adjust the vertices in ShoeSideCutout until the spline looks like the one in Figure 4-29. You might have to adjust the Bezier handles on some of the splines.

Figure 4-29. Adjusted side view profile for cutout

### Step 10. Path for cutout

The cutout object will be shorter than the outer shoe, so a new path must be made for lofting and deforming the cutout.

Under the Create panel, click on Shapes and choose Line. Under the Creation Type rollout, make sure Corner is turned on. Between the top and side profiles, create a straight line the same length as the cutout profiles. Name the path InnerPath.

Once the line is created, go to the Modify panel. Choose Edit Spline and turn on Sub-Object. Under the Edit Vertex rollout, click on Refine. Add three vertices to the path in the same way you added them to the outer shoe path.

> **TIP** *The length of the deformed object will be determined by the length of the path, not by the length of the fit shapes.*

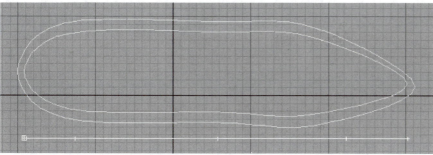

*Figure 4-30. Extra vertices on InnerPath*

Right-click on each vertex to change it to Corner type. Click on Sub-Object to turn it off.

### Step 11. Loft and deform the cutout

Now you have all the pieces you need to loft the cutout object. This object will be lofted, deformed and adjusted in the same way the outer shoe was created.

Select InnerPath. Under the Create menu, click the Geometry button. Select Loft Object from the pulldown list.

Under the Object Type rollout, click the Loft button. Under the Creation Method rollout, click Get Shape. Click on the ellipse. Under the Skin Parameters rollout, turn on the Skin checkbox. The lofted object appears in the viewport.

Click the Min/Max Toggle to go back to the four viewport display. Click Zoom Extents All.

Under the Modify panel, expand the Deformations rollout. Click on the Fit button. The Fit Deformation window appears.

Turn off the Make Symmetrical button. Click on Get Shape, and click on ShoeTopCutout in any viewport. Click the Zoom Extents button at the bottom of the Fit Deformation window.

Click on Display Y Axis. Click on ShoeSideCutout in any viewport.

Close the Fit Deformation window. Change the name of the new object to InnerShoe.

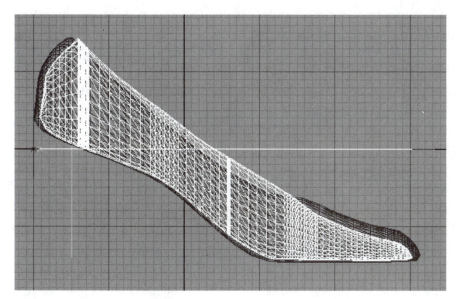

*Figure 4-31. Shoe cutout object*

Next you can redistribute the faces along the shoe in the same way you did for OuterShoe. With the object selected, go to the Modify panel. Expand the Skin Parameters rollout. Turn off the Adaptive Path Steps checkbox. Change Path Steps to 8, and change Shape Steps to 8.

*If both the outer shoe and the cutout are the same color, change the color of one to a contrasting color. This will make it easier to position the cutout in the shoe. To change an object's color, select the object and click on the color swatch next to the object's name in the panel.*

## Step 12. Clean the clutter

The next step is to move the cutout to the correct position for the boolean subtraction operation. But before you do this, do a little housekeeping on the screen.

Select all objects except OuterShoe and InnerShoe. There are several ways to select these objects. The easiest way is to click

Select by Name. Click on the first object you want to select. The object name is highlighted. Hold down the <Ctrl> key and click on the other objects you want to select. Click on OK to exit.

Under the Display panel, locate the Hide by Selection rollout. Click on Hide Selected. The selected objects are hidden from view. You can bring them back at any time by clicking Unhide All.

## Step 13. Align cutout with outer shoe

*Even though you selected and hid all objects except OuterShoe and InnerShoe, there are still two paths and shapes on the screen. The path and shape are an integral part of the lofted object. If you delete or hide the paths or shapes, the lofted objects will go with them.*

Now that you can see the objects more clearly, you can line up the cutout with the outer shoe in preparation for the boolean subtraction.

Click in the Front viewport. Choose Select and Move and turn on Restrict to Y. Move InnerShoe up by a few units so its sole sits inside OuterShoe, as shown in Figure 4-32.

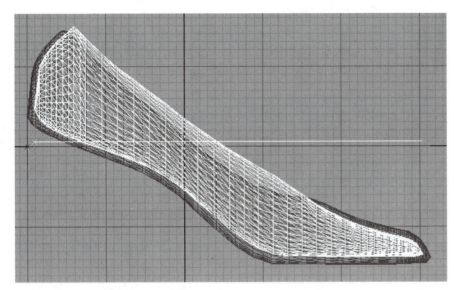

*Figure 4-32. InnerShoe position*

Click in the Top viewport. Turn on Restrict to Y 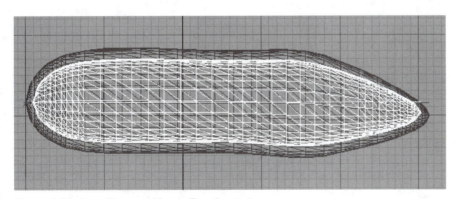 if neces-
sary. Move InnerShoe so it sits at the exact center of OuterShoe.
Zoom in to check that the cutout is as close to the center of
OuterShoe as you can get it.

*Figure 4-33. InnerShoe position in Top viewport*

> **TIP** *It is very important that you save the file at this point. Not only is it a good practice to save your work just before a boolean operation, but you'll need one of the objects in this file later on.*

> **TIP** *The boolean operation might result in both operands disappearing. If this happens, reload SHOE01.MAX. Move one of the objects slightly, save the file and try the boolean again. For a detailed explanation of why booleans sometimes don't work, see Chapter 2, Boolean Techniques.*

## Step 14. Boolean subtraction

You're now ready to perform the boolean subtraction. This is a good time to save your work. Save the file as SHOE01.MAX.

Select OuterShoe. Under the Create menu, choose Compound Objects from the pulldown list. Click on Boolean.

Note that under the Parameters rollout, Subtraction (A-B) is on by default. You have already chosen operand A by selecting OuterShoe before you started the boolean operation. Under the Pick Boolean rollout, turn on Instance. Click on Pick Operand B and select InnerShoe. Wait a few moments while MAX performs the boolean subtraction.

When the boolean operation is complete, the cursor appears. Other than that, the model looks the same. You can't see the hollowed-out shoe until you hide the InnerShoe object.

Select InnerShoe. Under the Display panel, click on Hide Selected. The hollowed-out shoe appears. Render the Perspective view for a look at the shoe.

> **TIP** Because you chose Instance before you did the boolean operation, operand B, the object InnerShoe, remained in the scene. This allows you to later change InnerShoe and have the boolean object automatically adjust to the change. If you had chosen Move instead of Instance, operand B would have disappeared from the scene.

*Figure 4-34. Booleaned shoe*

Note that the paths and shapes used to define the objects has disappeared. InnerShoe and OuterShoe are no longer defined as loft objects. OuterShoe is now a boolean object and InnerShoe is a mesh, so the paths and shapes are no longer used to define them.

Save your work as SHOE02.MAX.

## Tutorial 10
## Heel

The heel will be created by extruding and tapering a shape, then performing a boolean to make the heel fit the shoe perfectly.

### Step 1. Extrude the heel

Under the Create panel, choose Shapes. Click on Circle. In the Top viewport, draw a circle near the heel end of the shoe.

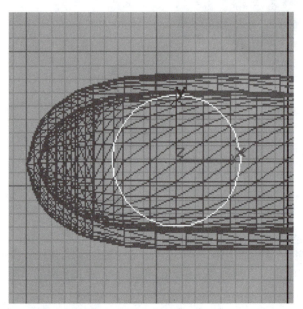

*Figure 4-35. Circle for heel shape*

Under the Modify panel, choose Edit Spline. Turn on Sub-Object and make sure Vertex is selected as the Sub-Object level. Select the rightmost vertex on the circle and delete it by pressing the <Delete> key on your keyboard.

Click Select and Move 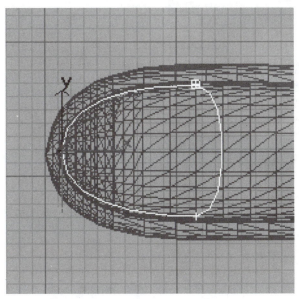 and turn on Restrict to X. Select the leftmost vertex on the circle. Move the vertex to the left until it sits just inside the rim of the shoe. Turn on Restrict to Y

and adjust the Bezier handles so the shape follows the shape of the shoe.

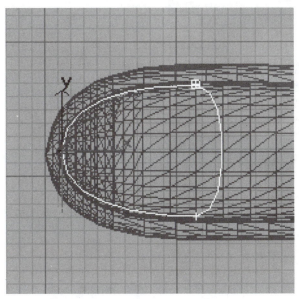

Figure 4-36. Adjusted heel shape

Change the Sub-Object level to Segment. Right-click on the rightmost segment and choose Line from the list. The segment changes to a straight line.

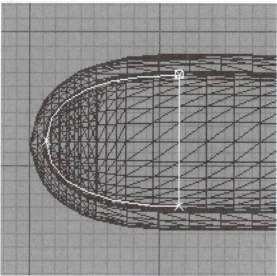

*Figure 4-37. Heel shape with straight segment*

Turn off Sub-Object. In the Front viewport, move the heel shape upward until it sits just inside the shoe.

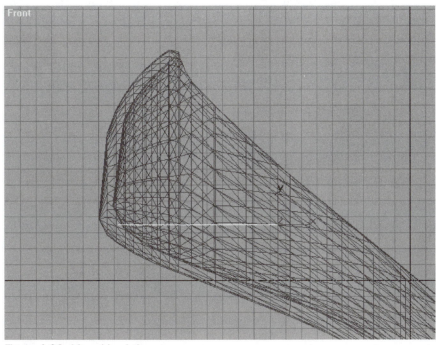

*Figure 4-38. Moved heel shape*

Under the Modify panel, choose Extrude. Under the Parameters rollout, change the Amount spinner so the heel extrudes downward. Extrude the heel until the bottom of the heel is even with the bottom of the shoe.

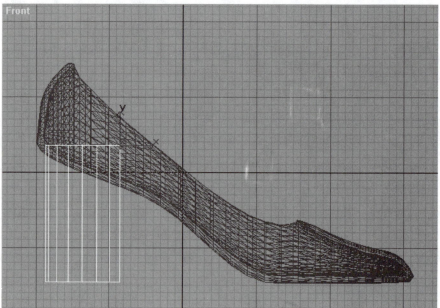

Figure 4-39. Extruded heel

Change the name of the object to Heel.

You could stop here and have a chunky retro shoe, but instead, you'll taper the heel for a more classic look.

## Step 2. Taper the heel

Click on Taper. Under the Parameters rollout, change the Amount spinner to taper the heel, as shown below.

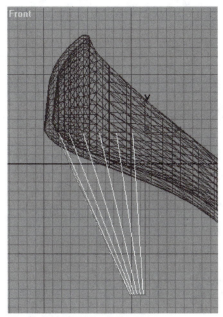

*Figure 4-40. Tapered heel*

The heel is a little too angled. Such a shoe would be difficult to stand on. The taper needs to be angled more to the left. This can be accomplished by changing the taper gizmo. The taper gizmo is a sub-object that controls the direction of the taper.

Turn on Sub-Object. The Gizmo level is selected. Choose Select and Move 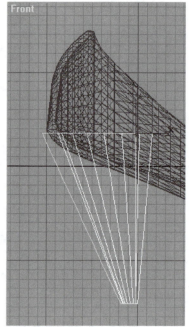 and turn on Restrict to X. Move the gizmo to the left in the Front viewport so the heel is angled more to the left.

*Figure 4-41. Gizmo moved*

Turn off Sub-Object.

### Step 3. Boolean heel

The heel is in place, but the top of the heel is visible inside the shoe. You can fix this problem by using a boolean operation to subtract the top part of the heel.

The shoe in its current state won't cut off the top of the heel. To boolean the heel, you'll need the original OuterShoe object.

Select the shoe. Rename the object to Shoe.

From the File menu, choose Merge. Pick the file SHOE01.MAX. Choose the object OuterShoe from the list. The original OuterShoe object appears.

Now you're ready to boolean the heel. This is a good time to save your work. Save the file as SHOE03.MAX.

Select the heel. Under the Create panel, click the Geometry button. Choose Compound Objects from the pulldown list. Click on Boolean. Turn on Move. Under the Parameters rollout, make sure Subtraction (A-B) is selected. Click on Pick Operand B. Press the H key and pick OuterShoe from the list.

In a few moments, a new heel appears. The heel fits the shoe perfectly.

Save your work as SHOE04.MAX.

## Tutorial 11
## Compound Materials

A shoe usually has one material on the inside and another on the outside. If you want to apply more than one material to an object, as in the case of the shoe, you'll need to use a non-standard material.

In MAX, materials are controlled by Material IDs. A Material ID is a number telling MAX which material should be assigned to a particular face. If you always use Standard materials, you've never had to work with Material IDs. A Standard material is assigned to an entire object without regard for its Material IDs.

Material IDs work in conjunction with a non-Standard type of material, the Multi/Sub-Object material. A Multi/Sub-Object material is comprised of several sub-materials, each with a number. The sub-material used for a face is the sub-material matching the face's Material ID.

Right now, the faces of both OuterShoe and InnerShoe are all assigned Material ID 1. When you perform a boolean subtraction, the faces created by the cutout take on the cutout's Material ID. If you change the Material ID of InnerShoe's faces to Material ID 2, the inside of the shoe can be assigned a material different from the outside.

### Step 1. Change Material ID

Under the Display panel, choose Unhide by Name. Pick InnerShoe from the list.

Select InnerShoe. Under the Modify panel, choose Edit Mesh. Click on Sub-Object and choose Face from the pulldown list.

This selection contains many rollouts. Go to the last rollout, Edit Surface. It will be easier to do this if you first close the Edit Face rollout.

Click on Select by ID. Make sure ID 1 is entered, and click on OK. All the faces in Innershoe turn red. You have just selected all the faces in Innershoe.

On the rollout, change ID to 2. Close the rollout. The faces on InnerShoe are now assigned Material ID 2. This means that the faces on the Shoe object which were cut out by InnerShoe also have Material ID 2.

Go back to the top of the panel and turn off Sub-Object. Under the Display panel, choose Hide Selected to hide InnerShoe.

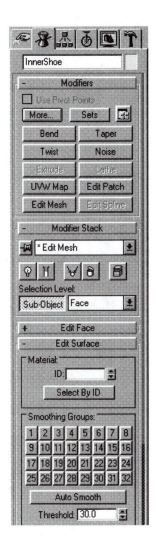

### Step 2. Create Material

Select both the shoe and the heel. Make sure the Perspective viewport is still set to Smooth and Highlight. If not, right-click on the Perspective viewport label to change the display.

Click on the Material Editor button  to open the Material Editor window. The sample box at the upper left corner, which shows a dark red material, is selected. Click on Assign Material

to Selection ![Assign Material to Selection icon] to put the red material on the shoe and heel.

Next to Type, click on the button labeled Standard. A selection list appears. Choose Multi/Sub-Object. A dialog box appears.

![Replace Material dialog box]

Choose to keep the old material. Click on OK. A new set of rollouts appear for the new material type.

Material #2 defaults to a gray material. Look at the shoe in the Perspective viewport. The inside of the shoe is gray. This is because the faces inside the shoe have been assigned Material ID 2. All faces with Material ID 2 take Material #2 from a Multi/Sub-Object material.

Under the Parameters rollout, click on Set Number. Set the number of materials to 2.

Above the Parameters rollout, change the name of Material #1 to Shoe Material. On the Parameters rollout, click on the button labeled Material #1 (Standard). The material name is Material

#1. Enter the name Outer Shoe. Click on Go to Parent.

Next to Material #2, Click on the button labeled (Standard). Change the name of the material to Inner Shoe. Click on Go to

Parent.

You now have a parent material called Shoe Material with two sub-materials, Outer Shoe and Inner Shoe.

Click on the label for Material #2, Inner Shoe. Change the Diffuse color to a pale yellow. Note that the inside of the shoe in the Perspective viewport changes accordingly.

You can now change the sub-materials in any way you like to get the colors you want both inside and outside the shoe.

Close Material Editor window when done. Save your work as SHOE05.MAX.

### Things to Try

The tip of the heel on a ladies' shoe is often covered with a protective black cap. To put on a cap, loft the heel with six or seven segments, and perform the taper and boolean as described in this chapter. Select the faces at the bottom of the heel and assign them another Material ID. You can then make a black sub-material to assign to this part of the model.

The instep of the shoe is often a different color from the rest of the inside. To simulate this, you can select faces from the object Innershoe and assign them a different Material ID, or you can make a separate object from the faces and raise it up to simulate a real instep. This same object can also be copied down just under the shoe to make a sole.

A high heeled shoe often has stitching or trim around the upper edge of the shoe. To simulate this, carefully construct a loft path around the upper edge of the shoe, and loft a very small circle on the path. Make the lofted trim the same color as the shoe, or use a contrasting color. This technique was used to create the trim on the shoe in the file HIGHSHOE.MAX, which can be found in the \MICHELE\MESHES directory on the CDROM. This file was used to make the image at the beginning of this chapter.

# c h a p t e r

## Patchwork Quilt

In your work with MAX, there may have been times when you wish you could push or pull a portion of the model and sculpt the mesh like taffy. Many of MAX's modeling tools, such as Bend, Taper and Edit Mesh, are fine for creating man-made solid objects like spaceships, buildings and teacups. But there are many objects that are difficult to create with these tools alone.

This is where Bezier patches come in. To understand a Bezier patch, consider a Bezier curve on a 2D shape. The curve is controlled by handles. As long as the handles are kept in a straight line, the curve remains smooth.

The Edit Patch modifier temporarily converts the surface of an object into a series of Bezier patches. When you move the patch handles, the surface of the object is reshaped smoothly, the same way a 2D shape remains smooth when you move the handles of a Bezier curve.

*Figure 5-1. Patchwork quilt created with Bezier patches*

In this chapter, you'll use patches to mold a flat grid into a quilt to fit a bed.

The only way to really understand patches is to experiment with them. The instructions here and in the next tutorial are designed to familiarize you with patches and their uses.

## Working with Patches

To learn more about patches, we'll use a cylinder with an Edit Patch modifier.

Create a cylinder of any size in any viewport. Click Zoom Extents All. Change the Perspective viewport to Smooth + Highlight. Click Arc Rotate. Rotate the Perspective view to look down on the top of the cylinder, like Figure 5-2.

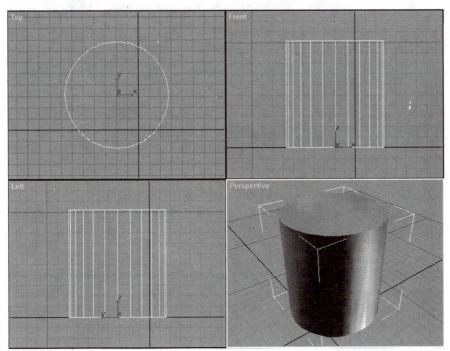

Figure 5-2. Cylinder

Under the Modify panel, click Edit Patch. A lattice appears around the cylinder.

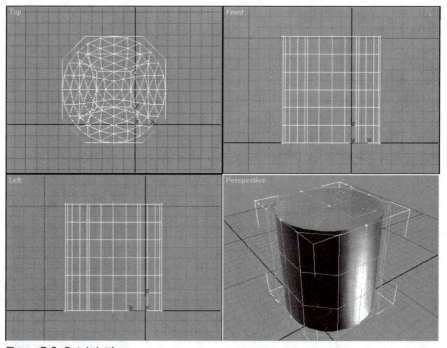

Figure 5-3. Patch lattice

The lattice is used to control the patches. In the case of the cylinder, the lattice sits outside the cylinder and is easily visible.

The Vertex sub-object level is automatically selected when you apply an Edit Patch modifier.

Click Select and Move. 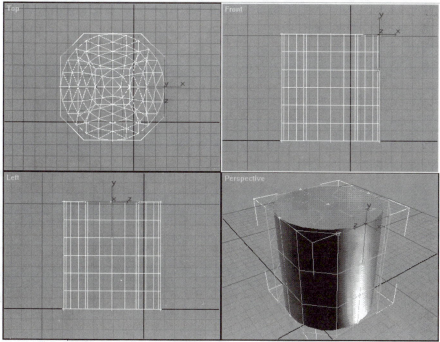 In the Front viewport, click the vertex at the upper right of the cylinder. Four handles appear around the vertex.

*Figure 5-4. Selected vertex and handles*

In the Front viewport, pull the vertex up and to the right. When you release the vertex, note the change in the object in the Perspective viewport.

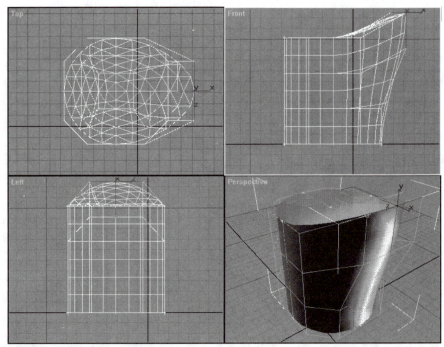

Figure 5-5. Modified cylinder

The cylinder remains smooth.

Click Select and Rotate. In the Front viewport, rotate the vertex by about 60 degrees.

**Take care to click on the vertex itself and not one of the handles when rotating the vertex. Handles appear as green boxes, while a vertex is represented by a red dot inside a green box.**

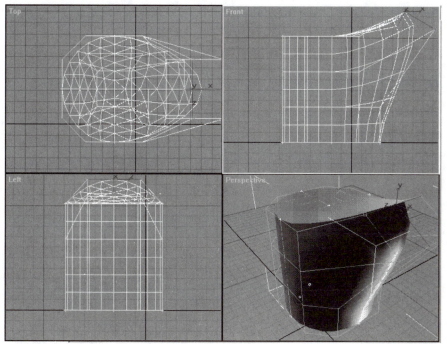

Figure 5-6. Selected vertex and handles

No matter what you do to the vertex, the object is deformed smoothly.

In the Left viewport, move your cursor to the bottom center of the cylinder until the cursor turns into a selection cursor. Click to select the vertex.

In the Top viewport, you can see that the vertex has four handles arranged in the same plane. Compare this handle arrangement with the handles on a vertex on the edge of the cylinder. The Bezier handles on a patch vertex are arranged differently depending on the part of the object they correspond to.

Practice pushing and pulling vertices on the patch lattice. Try moving and rotating individual handles to see the effect on the object. If you turn on the Lock Handles checkbox under the Edit Vertex rollout, the relative angles of the handles are maintained during all modifications.

Select groups of vertices and move them all at the same time. You can move and rotate groups of vertices, but you can only move one handle at a time.

You can also try working with the other sub-object levels, Edge and Patch. In general, the Vertex sub-object level works the most intuitively.

When you've finished experimenting with the cylinder, create a box of any size in any viewport. Apply an Edit Patch modifier to the box. The lattice sits right on the surface of the object.

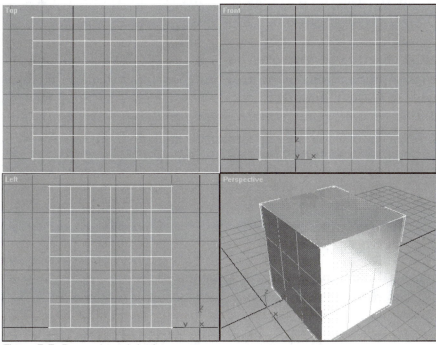

*Figure 5-7. Box and patch lattice*

Select different vertices on the box lattice to see how their handles are arranged.

Continue experimenting until you feel comfortable working with patches, then go on to the next tutorial.

# Tutorial 12
# Building a Patchwork Quilt

In this tutorial you'll use a patch grid to create a smoothly draped object, a quilt in the shape of a bed.

A patch grid is a grid with built-in patch information. You can make modifications to the grid using the patches, and the object will always stay smooth.

Under the File menu, choose Reset. Save your previous work if you like. Answer Yes when asked if you really want to reset.

Change the Perspective viewport to Smooth + Highlight.

### Step 1. Create patch grid

Under the Create panel, click Geometry. Choose Patch Grid from the pulldown menu.

Under the Object Type rollout, click on Quad Patch. In the Top viewport, click and drag to draw a grid of any size. Under the Parameters rollout, set the following values:

| | |
|---|---|
| **Length** | **100** |
| **Width** | **100** |
| **Length Segs** | **3** |
| **Width Segs** | **3** |

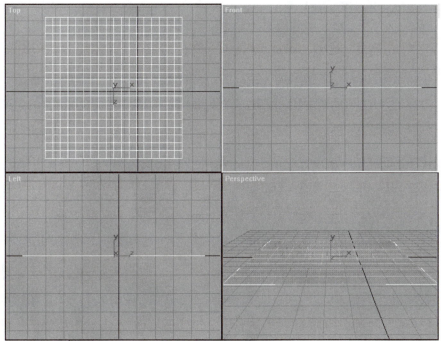

*Figure 5-8. Quad patch*

Name the object Quilt.

### Step 2. Shape quilt

Go to the Modify panel. Click Edit Patch. Turn on Sub-Object and choose Vertex as the sub-object level.

Under the Edit Vertex rollout, turn on the Lock Handles checkbox.

In the Top viewport, select the vertices along the outside edge of the quilt, as shown in Figure 5-9. Use the <Ctrl> key when selecting to add to your selection set. Use <Alt> to subtract from the selection set.

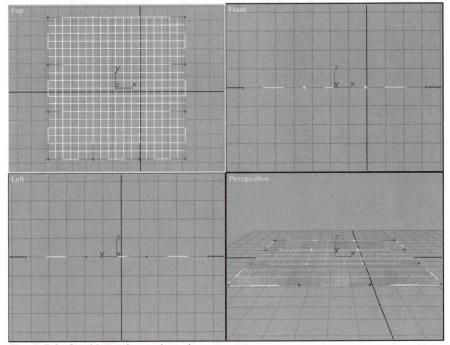

*Figure 5-9. Outside vertices selected*

Activate the Front viewport and click Select and Move. 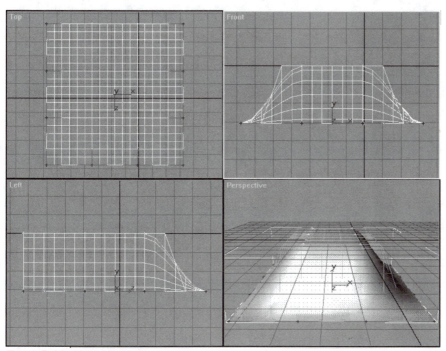 Turn on Restrict to Y. ▢ Move the selected vertices downward by about 30 units. Click Zoom Extents All. ▢

*Vertices will move or rotate as a group, but handles can only be moved individually.*

*Figure 5-10. Moved vertices*

Look at the patch grid in the Perspective viewport. The mesh is still smooth despite the changes to the patch grid.

In the Top viewport, select just the side vertices, as shown in Figure 5-11.

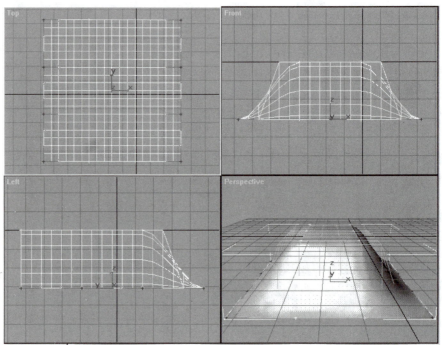

Figure 5-11. Side vertices selected

Activate the Top viewport. Click Select and Non-uniform Scale ![icon] and turn on Restrict to X. ![icon] In the Top viewport, scale the vertices to about 55%.

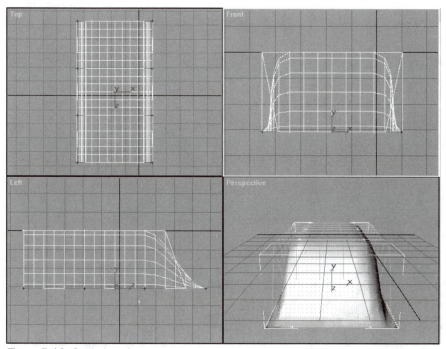

*Figure 5-12. Scaled vertices*

The sides of the quilt are in position. Now you'll move the end of the quilt.

Select just the vertices at the bottom of the quilt in the Top viewport. Turn on Restrict to Y 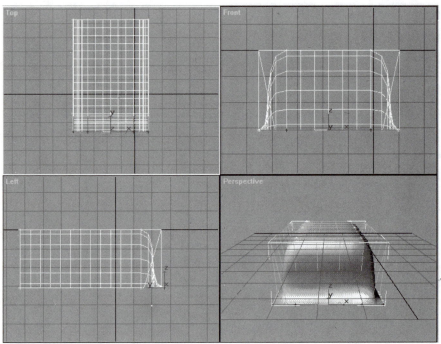 and move the vertices upward by about 20 units.

Figure 5-13. Bottom of quilt moved

In the Perspective viewport, you can see that the quilt is starting to take shape. However, the vertex handles all along the edge are still oriented in the same plane as the original grid. The vertices all around the edges must be rotated to make the quilt look as if it's hanging from a bed.

## Step 3. Rotate vertices

The edges of the quilt are currently a little curled. This is because the patch vertex handles are still pointing inward. The handles must be rotated to make the quilt hang correctly.

The corner vertices will be treated separately from the rest of the quilt edge. In this step, make sure the corner vertices are not part of any selection set you rotate.

Select the vertices at the left edge of the quilt. In the Top viewport, deselect the vertex at the lower left corner.

Activate the Front viewport and click Select and Rotate. Make sure Restrict to Z is on. Rotate the selected vertices by about 80 degrees.

Activate the Perspective viewport and click Arc Rotate. 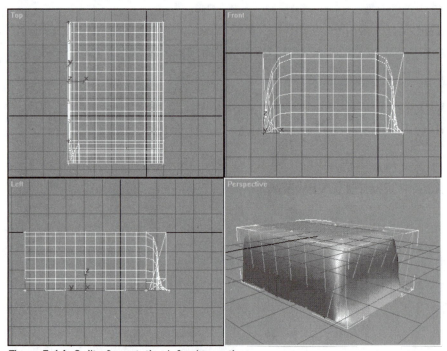 Rotate the viewport so you can see how rotating the vertices affected the mesh.

Figure 5-14. Quilt after rotating left edge vertices

Now do the same for the other side of the quilt. Select the vertices at the lower right of the quilt and deselect the corner vertex. In the Front viewport, rotate the vertices by about -80 degrees.

In the Top viewport, select all the vertices at the bottom of the quilt except the corner vertices. Activate the Left viewport. Rotate the selected vertices by about -80 degrees.

The sides of the quilt now hang down straight.

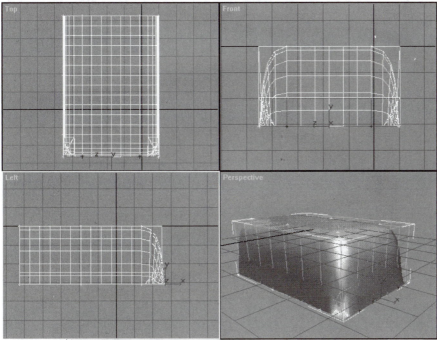

*Figure 5-15. Quilt after rotating edge vertices*

### Step 4. Corner vertices

The corner vertices must be pulled down and rotated to simulate the corners of a quilt hanging from a bed.

Use the Top viewport to select both bottom corner vertices. In the Front viewport, move both vertices downward by about 10 units.

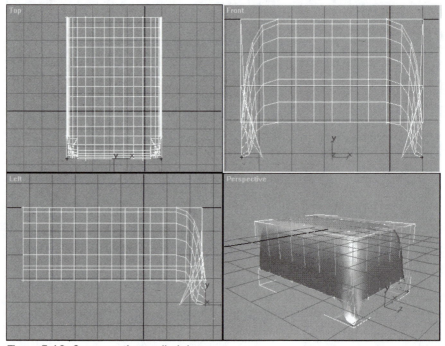

*Figure 5-16. Corner vertices pulled down*

With both corner vertices still selected, choose Select and Rotate. Activate the Left viewport. Rotate both vertices by -45 degrees.

In the Front viewport, select just the left corner vertex. Rotate the vertex by about -45 degrees. Select the right corner vertex and rotate it by about 45 degrees.

Choose Select and Uniform Scale. 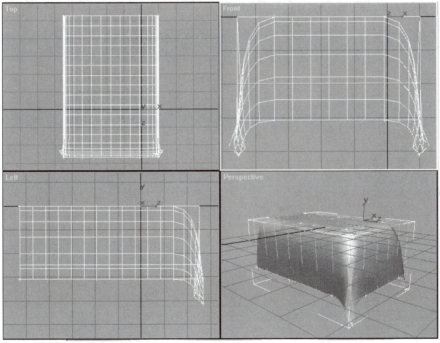 Scale each corner vertex individually to about 50%. This scales the handles to make the corners look more like a real quilt.

*Figure 5-17. Quad patch corners*

### Step 5. Pillows

The other end of the quilt must be puffed up a little to simulate pillows underneath the bedspread.

In the Top viewport, select the two top center patch vertices. In the Left viewport, move the vertices up by about 8 units.

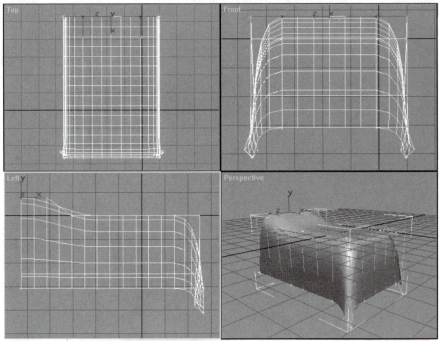

Figure 5-18. Top vertices moved up

In the Left viewport, pull the rightmost handles of each vertex up and to the right. The handles of each vertex must be moved individually.

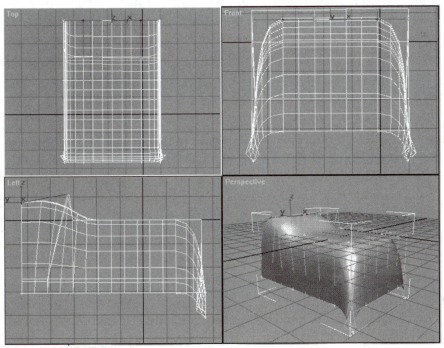

*Figure 5-19. Finished quilt*

The quilt is now complete. Save your work as QUILT01.MAX.

Feel free to work with the vertices to create your own unique bed-spread. Be sure to save your work when finished.

## Tutorial 13
## Mapping Patches

Using mapping coordinates with the patch grid quilt created in the previous tutorial presents an interesting problem. Ideally, you would use an Edit Mesh modifier to assign different material IDs to different faces and use a Multi/Sub-Object material to make a patchwork quilt. However, the quilt is now reshaped and it would be difficult to pick faces from all its sides with accuracy.

However, if you had picked the faces before reshaping the quilt, the Edit Mesh modifier would have destroyed the original patch information, making it impossible to shape the quilt so easily.

To illustrate this problem, reset MAX and create another patch grid of any size. Apply an Edit Mesh modifier. Now choose Edit Patch and turn on the Vertex Sub-Object level. Select a vertex and move it around.

The big patches are now small patches, one for each face. When you move a vertex, the object deforms with hard angles, not the smooth shapes that patches are usually so good for.

To get around this problem, you'll temporarily turn off the Edit Patch modifier. This will display the quilt as a flat quad patch, which will allow you to pick faces for the patchwork quilt material.

Before beginning, load the file QUILT01.MAX from the previous tutorial.

## Step 1. Turn off Edit Patch modifier

Under the Modify panel, click on the Active/inactive modifier

toggle. 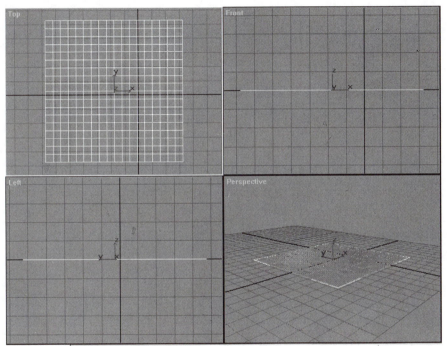 This turns off the Edit Patch modifier. The quilt appears as a flat grid.

> **If you want to apply further modifiers to the quilt, be sure to go back up to the top of the stack first. Otherwise, any new modifiers will be applied between the Edit Patch and Edit Mesh modifiers.**

*Figure 5-20. Quilt with Edit Patch modifier turned off*

You'll turn the Edit Patch modifier back on later, after you've assigned Material IDs.

## Step 2. Assign Material IDs

Next you'll apply an Edit Mesh modifier to the object so you can pick faces for different Material IDs. Click Edit Mesh. Select Face as the sub-object level. Collapse the Edit Face rollout so you can see the top of the Edit Surface rollout.

Select several faces on the patch grid at random. When you've selected about 20 faces, change the Material ID setting in the Edit Surface rollout to 2. This assigns Material ID 2 to the selected faces.

Select another series of faces at random. Assign these faces Material ID 3. Select more faces and assign them Material ID 4.

## Step 3. Multi/Sub-Object Material

*The quad patch quilt is a one-sided object with normals pointing upward. If you want to see the underside of the quilt in the rendering, you must apply a 2-sided material or render with the Force 2-Sided checkbox turned on.*

Open the Material Editor. Create a Multi/Sub-Object material with four sub-materials. To do this, click on the button labeled Standard just under and to the right of the sample boxes. Choose Multi/Sub-Object. Set the number of materials to 4. Create four separate materials, one for each Material ID.

If you like, use the QUILT*.TIF bitmaps from the \MICHELE\MAPS directory on the CDROM. These are tiling bitmaps that can be used as Diffuse maps.

The file QUILTBMP.TIF is a bump map that can be used to simulate the stitching around each square on a quilt. Set the tiling of the bump map to 18x18 to match the tiling of the grid faces.

Click Assign Material to Selection to apply the material to the quilt object. Each of the four materials will be assigned to faces with the matching Material ID number. Close the Material Editor.

If you're not happy with the distribution of the Material IDs, continue to select faces and assign them to different Material IDs until you like the looks of the quilt.

### Step 4. Restore Edit Patch modifier

Under the Modifier Stack rollout, click the small down arrow to display the modifier stack. Pick Edit Patch. Click the Active/inactive modifier toggle to make the Edit Patch modifier active again. The quilt is restored to its earlier shape.

Save your work as QUILT02.MAX.

A finished brass bed with a larger quilt patch grid can be found in the \MICHELE\MESHES directory in the file BED.MAX on the CDROM. Feel free to remove the quilt and replace it with your own. You can see the rendered image by viewing the file BED.TGA in the \MICHELE\IMAGES directory.

# c h a p t e r

6

## Lighting Effects

With 3D Studio MAX, it's possible to create a number of lighting effects. Many of MAX's lighting tools are surprisingly easy to use and can make stunning effects with just a few steps. Volume light, for example, can create a dense, complete look in an otherwise simple model.

In the tutorials in this chapter you'll learn how to use volume light and other techniques to make impressive scenes in a short time.

## Tutorial 14
## Starburst

One of MAX's best features is its ability to create realistic outer space backgrounds and scenes with very little effort. In this tutorial, you'll make a burning star against a starfield background with noise and volume light effects.

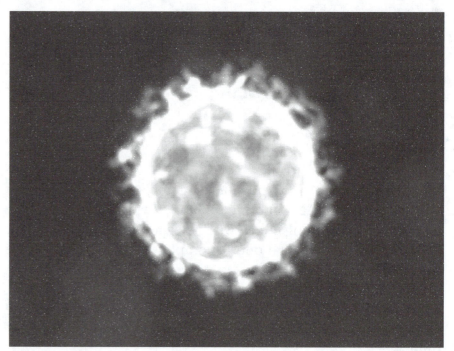

*Figure 6-1. Burning star*

To see an animation file created with this technique, view the file SUNBURST.AVI in the \FRANK\AVI directory on the CDROM.

## Step 1. Background

To set up the starfield background, you'll use a noise effect.

Click Material Editor. 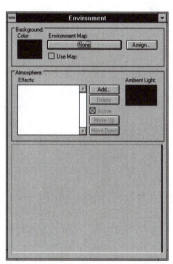 You'll be using the Material Editor to set up the background.

Under the Rendering menu, choose Environment. The Environment dialog box appears.

*Figure 6-2. Environment dialog box*

On the Environment dialog box, click the Assign button. Pick Noise from the list that appears and click OK to return to the Environment dialog box.

The button under Environment map is now labeled Map #1 (Noise). Click this button. The Put to Material Editor dialog box appears.

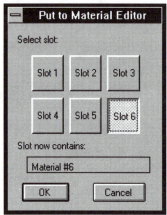

Figure 6-3. Put to Material Editor dialog box

This dialog box allows you to use a sample slot in the Material Editor to set up the environment.

Click Slot #6 and click OK to exit the Put to Material Editor dialog box. Exit the Environment dialog box.

In the Material Editor, the sixth sample slot now holds the noise material. Click on the sixth sample slot. The rollouts in the Material Editor change to settings for the noise effect.

Under the Noise Parameters rollout, change Size to 0.1 and Low to 0.65. Change Color 1 to a deep blue color of your choice.

*TIP* *The environment effect will not appear in a rendering of an orthographic viewport.*

Render the Perspective view. The view shows a starfield against a blue background.

## Step 2. Nebula

The background is okay, but would benefit from some color variations to suggest a nebula. These color variations will be added with more noise effects.

MAX allows you to apply sub-levels of noise to a material. For example, the Color 1 setting for the noise material could be another noise material, and Color 2 could be another one. These noise materials would also have Color 1 and Color 2 settings which could in turn hold more noise settings. With this technique, you can create varied surfaces and backgrounds with smooth color transitions.

This technique will be used in this step to make the nebula background.

Sample slot 6 should still be selected. Click the button labeled None across from Color 1. Select Noise from the list.

The rollouts that appear apply to this new Noise effect. Under the Noise Parameters rollout, turn on Fractal. Set Low to 0.4.

Change Color 2 to a dark blue with the following settings.

| | |
|---|---|
| Red | 22 |
| Green | 33 |
| Blue | 117 |

Render the Perspective view. Some color variation has been added. To add more, you'll add another Noise effect.

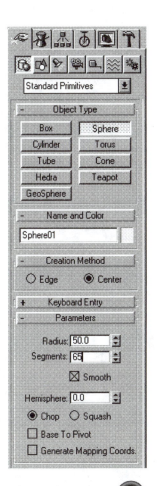

Click on the button labeled None across from Color 1. Under the Noise Parameters rollout, turn on Fractal. Set Low to 0.4 and Phase to 7.3. Change Color 2 to a deep magenta with the following settings.

**Red**    77
**Green**  0
**Blue**   76

Close the Material Editor. Render the Perspective viewport. The nebula background is complete.

**Step 3. Star**

The star will be made with a simple sphere.

Under the Create panel, click Geometry. Click Sphere. Create a sphere in any viewport. Under the Parameters rollout, change Radius to 50 and Segments to 65. Change the name of the sphere to Star.

Click Zoom Extents All.

Now you need an appropriate material for the star.

Click Material Editor. Select the first sample slot. Under the Basic Parameters rollout, set Self-Illumination to 100. Click on the small box next to the Diffuse color swatch. Choose Noise from the list.

Under the Noise Paramaters rollout, turn on Fractal. Set Size to 10 and High to 0.71.

Change Color 1 to a bright yellow with the following settings.

**Red**    246
**Green**  255
**Blue**   0

Click the button labeled None across from Color 2. Select Noise from the list. Under the Noise Parameters rollout, turn on Fractal. Set Size to 10 and Low to 0.525.

Set Color 1 to a bright orange with the following settings.

**Red**    255
**Green**  74
**Blue**   0

Set Color 2 to a dark orange with the following settings.

**Red**    59
**Green**  44
**Blue**   0

Select the sphere. Click Assign Material to Selection to assign the material to the sphere.

Render the Perspective view to see the sphere against the background.

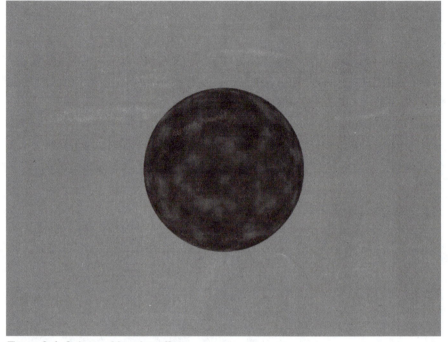

*Figure 6-4. Sphere with noise effects*

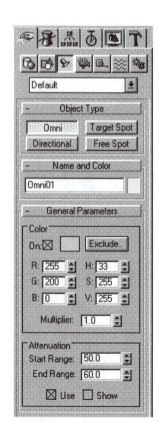

## Step 4. Omni lights

To use volume light effects, you first need some standard lights.

Under the Create panel, click Lights. Click Omni. Place an omni light in the center of the sphere. Under the General Parameters rollout, change the color to a bright yellow with the following settings.

| | |
|---|---|
| **Red** | **255** |
| **Green** | **200** |
| **Blue** | **0** |

Under the Attenuation section, turn on the Use checkbox. Change Start Range to 50 and End Range to 60. Name the light Glow.

As you change the Start Range and End Range, you can see circles in each viewport representing the ranges.

Attenuation causes the light to shine only within a certain range of the light source. In this case, the light will shine only in areas that are 50 to 60 units from the omni light. Remember that the sphere has a radius of 50 units. The light will shine from the outside of the sphere to the area 10 units out from the sphere.

Activate the Top viewport. Click Select and Move and turn on Restrict to X. Hold down the <Shift> key and click and drag on the omni light. Move the light outside the sphere. On the Clone Options dialog box, enter the name Flame.

This light will eventually be moved back into the sphere. The light has been moved outside the sphere so it will be easier to pick when setting up volume lighting.

Under the Modify panel, change the new light's parameters. Change Start Range to 60 and End Range to 80.

### Step 5. Volume lighting

Volume lighting is an effect that gives light the appearance of density. This effect can be used to simulate the light emitted by the burning gasses of a star.

A volume light is not a type of light such as an omni or spotlight. A volume light is actually an atmosphere effect applied to an ordinary light. In this step you'll apply a volume light effect to each omni light.

Under the Rendering menu, choose Environment. Under the Atmosphere section, click the Add button. A list of atmospheric effects appears. Choose Volume Light from the list.

The Environment dialog box changes to show rollouts for setting up the volume light effect. Under the Volume Light Paramters rollout, click Pick Light. Click on the light inside the sphere. The name Glow appears in the text field across from the Pick Light button to indicate that it has been selected.

In the Volume section of the Volume Light Parameters rollout, change Density to 30. Under the Noise section, turn on the Noise On checkbox. change Amount to 0.5.

*You must render the Perspective view and not an orthographic view to see the effect. A camera view will also render the effect. In addition, the Render Atmospheric Effects checkbox must be on in order for the effect to appear.*

Click the Add button again and choose Volume Light from the list. Click Pick Light and click on the Flame light. Change Density to 30. Turn on the Noise On checkbox and change Amount to 0.98. Change Uniformity to -0.1 and Size to 10.

Close the Environment dialog box. Move the Flame light back to the center of the sphere.

The volume lighting is now set up. Render the Perspective viewport.

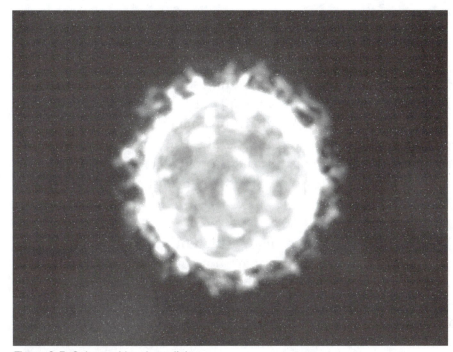

*Figure 6-5. Sphere with volume light*

## Things to Try

The volume light settings can be easily animated. To try this out, turn on the Animate button. Move the frame slider to frame 100. Under the Rendering menu, choose Environment. Choose one of the Volume Light entries from the Effects list. Change any or all of the parameters by a small amount. Move the frame slider to frame 100 and change parameters again. When you slide the frame slider back and forth, you'll see the parameter values change. Render the animation to see the changes over time.

You can also make a scene with a solid planet hiding the burning star, then moving aside to reveal the star. View the animation file SUNBURST.AVI in the \FRANK\AVI directory on the CDROM for an example of this effect.

## Tutorial 15
## Laser Show

You may have seen the opening of a sports program featuring a laser light show, where images appear to be drawn against a surface with a laser beam. When the beam moves very fast and traces the same shape over and over again, the image becomes more and more evident. This tutorial presents techniques that can be used to simulate this type of laser effects show with 3D Studio MAX.

This tutorial deals with many different animation controllers, including expressions. If you are not comfortable assigning animation controllers to objects in 3D Studio MAX, please look over the section titled Using Expression Controllers in the 3D Studio MAX User's Guide.

You can also read over the beginning of Tutorial 17, where you will find a brief introduction to expression controllers.

## Step 1. Load file

Load the file LASERFX.MAX from the CDROM. This file can be found under the \KYLE\MESHES directory.

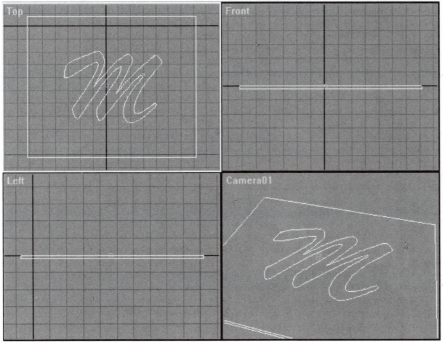

*Figure 6-6. File LASERFX.MAX*

This file contains a shape of a capital letter M as well as a small dummy object.

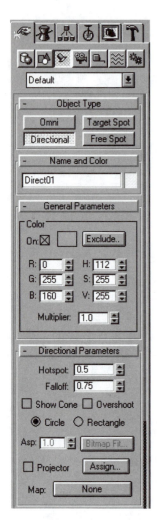

## Step 2. Create light

A directional light will be used for the laser beam.

By default, the hotspot and falloff angles of a directional light are always separated by 2 degrees. The falloff and hotspot must be tighter than this in order to make a laser beam. Before placing the lights, you'll change the default settings.

Under the File menu, choose Preferences. Click on the Rendering panel. Under HotSpot/Falloff, change the Angle Separation value to 0.0. Click on OK to save the change and exit the dialog box.

Under the Create panel, click Lights. Click on Directional. Place the light at the center of the M in the Top viewport.

Under the Directional Parameters rollout, change the light's HotSpot value to 0.5. Change Falloff to 0.75.

Under the Shadow Parameters rollout, turn on the Cast Shadows checkbox. Under the General Parameters rollout, set the light's color to the following.

**Red**      **0**
**Green**  **255**
**Blue**    **160**

Move the light to the position shown in Figure 6-7.

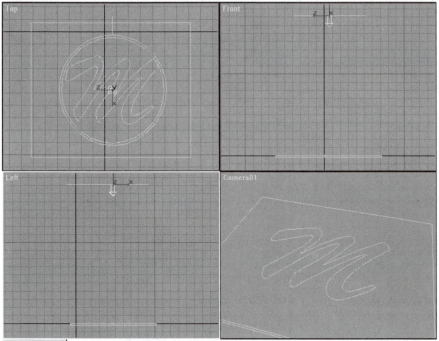

Figure 6-7. Placement for directional light

*A directional light works differently from an ordinary spotlight. A spotlight emits from a single point, where a directional light emits from an area. You might think of a directional light as a cylinder, with parallel rays of light. A spotlight is more like a cone. For this reason, directional light is better suited to simulate a laser beam.*

### Step 3. Set up volume light

In order to see the laser beam, you need to set up the light as a volume light. Volume lighting gives a light the appearance of density, such as search lights through fog or a beam of light through a smoky room.

Under the Rendering menu, choose Environment. Under Atmosphere, click Add. In the Add Atmospheric Effect dialog box, choose Volume Light and click OK.

Under the Volume Light Parameters rollout, click Pick Light. Select the directional light.

Set the volume light's settings as follows.

| | |
|---|---|
| **Density** | **100.00** |
| **Max Light** | **100.00** |
| **Auto checkbox** | **OFF** |
| **Sample Volume %** | **1** |

Close the Environment dialog box.

## Step 4. Make the dummy trace the M

In order to get the light to trace the M, it must be made to look at the M spline. This is accomplished by making the dummy object trace the M, then making the light look at the dummy.

Select Dummy01. Under the Motion panel, expand the Assign Controller rollout. Choose Position:Bezier Position from the list.

Click on the Assign Controller button. Choose Path from the list and click OK. Under the Path Parameters rollout, click Pick Path. Click on M-Spline in the Top or Camera viewport.

Dummy01 now traces the M-Spline.

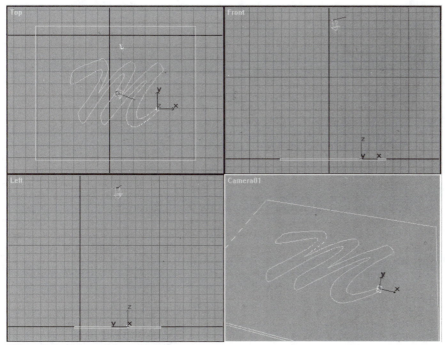

Figure 6-8. Dummy01 on M path

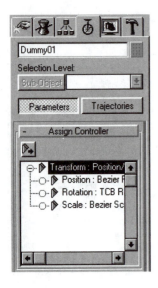

## Step 4. Light motion

Next you'll set up the light to look at the dummy object.

Select the light. Under the Motion panel, expand the Assign Controller rollout.

Choose Transform:Position/Rotation/Scale from the list. Click on the Assign Controller button.

Choose LookAt from the list and click OK. Under LookAt Parameters, click on Pick Target. Click on Dummy01.

Move the frame slider back and forth. The dummy traces the M, and the light always looks at the dummy.

## Step 6. Adjust animation

By default, when you assign a Path Controller to an object, it will trace the path once during the length of the animation. This is controlled by a parameter of the Path Controller named Percent.

Percent tells the object on what level of the path it should be at any given frame. At frame 0, the Percent value is set to 0.0, and at the last frame the Percent value is set to 100. This causes the object to travel once along the entire length of the path, traveling smoothly and regularly the whole way.

For this animation, the laser beam shouldn't move slowly, much less only once around the path. To make the object continuously loop around the path, its Parameter Curve Out-of-Range Types need to be modified.

Click on Track View 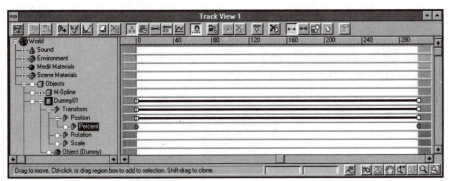 to open the Track View window. Select the Percent parameter of Dummy01 from the scene hierarchy list as shown below.

Figure 6-9. Track View with Percent selected

Click Function Curves. Click Parameter Curve Out-of-Range

Types. A dialog box appears.

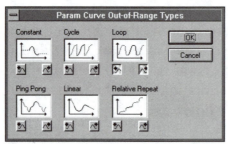

Figure 6-10. Parameter Curve Out-of-Range Types dialog box

Choose Loop and click OK. Now Dummy01 will continue to loop as long as the animation is played.

### Step 7. Change dummy loop time

Dummy01 is now looping once every 300 frames. A loop once every 100 frames is more desirable.

Click on the curve. This activates the curve and displays its two end points. Click on the rightmost endpoint, selecting it. Right-click on the same endpoint. A dialog box appears. Change the Time value to 100, as shown in Figure 6-11.

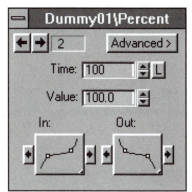

Figure 6-11. Changing the Time value for Percent

Exit from the dialog box. Now the Track View window looks like Figure 6-12. Can you see how the percent value is looping?

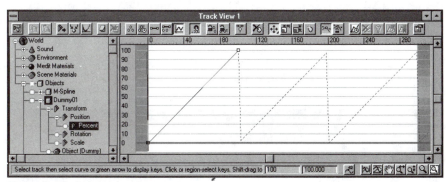

Figure 6-12. Percent looping

The laser beam should start out tracing the path slowly, then speed up. You could do this by editing keyframes individually, but it would be a long, arduous task. To make this task easier, you can apply an Ease Curve to the Path Percent controller of Dummy01. An Ease Curve affects the timing of a superior function curve. Please refer to the 3D Studio MAX User's Guide for a more in-depth discussion of Ease Curves.

Click on Apply Ease Curve on the Track View window. 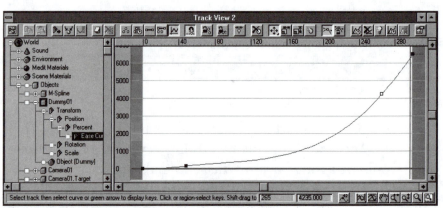 Expand Percent and highlight Ease Curve.

Click on the curve. Click Add Keys and click anywhere on the curve to add another control point.

Select each control point and change the values in the text entry box at the bottom of the Track View window for each point. Change the values as follows.

| Time | Value |
|------|-------|
| 0    | 0     |
| 48   | 154   |
| 265  | 4235  |
| 300  | 6500  |

The curve should now look like Figure 6-13.

Figure 6-13. Ease Curve applied to the Path Percent track

## Step 8. Motion blur

During approximately the last third of the animation, Dummy01 starts moving very fast. The rendered animation will look like random, flickering beams. The action can be made smoother with motion blur.

Under the Rendering menu, choose Video Post. The Video Post window appears.

Click on Add Scene Event. 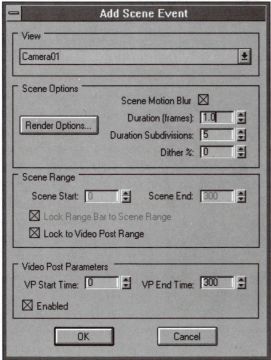 The Add Scene Event dialog box appears.

*Figure 6-14. Motion blur settings*

*Motion Blur is a technique that creates the illusion of very fast motion. It works by rendering multiple images in sequence and then compositing them together to form one frame. MAX composites each pass with an intensity based on how many Duration Subdivisions you specify.*

*If you render an image with three Duration Subdivisions, each Duration Subdivision is composited at 33.3%, or one-third intensity. Likewise, 10 Duration Subdivisions cause each pass to be composited at 10%, or one-tenth intensity.*

Turn on the Scene Motion Blur checkbox. Set the Scene Motion Blur settings as follows.

| | |
|---|---|
| **Duration (frames)** | 1.0 |
| **Duration Subdivisions** | 5 |
| **Dither %** | 0 |

Click OK to close the dialog box and save the settings.

### Step 9. Render sample frame

Next you'll do a quick render to see how the motion blur settings affect the scene.

Click Execute Sequence. 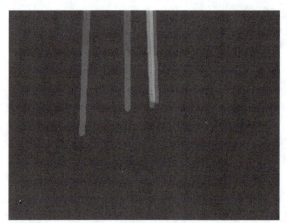 Specify Single Frame and type in 130. Choose an output size of 320x240 and click Render. Wait a few moments while MAX renders several images to create the effect of motion blur.

*Figure 6-15. Rendering of laser*

Close the Video Post window.

## Step 10. Create and offset dummies

The laser, though it has volume, doesn't show up very well. Adding more duration subdivisions won't solve the problem. Each subdivision is composited at a lesser intensity, so more subdivisions will only make the image darker.

The solution is to add more laser beams. However, this must be done with care. If you simply make a copy of the laser beam and render another test, it will look exactly the same as the previous render.

What you'll do is create copies of the laser beam, offsetting each new copy very slightly, thus making sure that each copy will not line up exactly with any other laser beam.

To offset each copy of the laser beam, you'll use a time unit called a *tick*. In MAX, a tick is 1/4800 of a second. A standard playback speed for MAX is 30 frames per second. This translates to 160 ticks per frame.

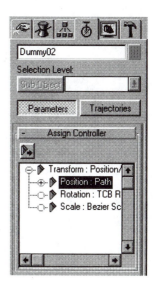

Select Dummy01. From the Edit menu, choose Clone. Specify Copy in the Clone Options dialog box and click OK.

Select Dummy02. Under the Motion panel, expand the Assign Controller rollout. Choose Position:Path from the list. Click on the Assign Controller button. Choose Position Expression from the dialog box and click OK.

Click on Position:Position Expression, then right-click on the selection and choose Properties. The Expression Controller: Dummy02 Position dialog box appears.

In the Create Variables section, type Dummy01Pos into the Name field and click on Vector. Change the Tick Offset to -15 and click on Create. Fifteen ticks is the equivalent of 1/320 of a second. This tick offset value will shift the laser beam copy from the original laser beam by 1/320 of a second.

Click on the Assign to Controller button. Choose Dummy01/Position:Path from the list and click OK.

Change the [0, 0, 0] in the Expression area to the following.

**[Dummy01Pos.x, Dummy01Pos.y, Dummy01Pos.z]**

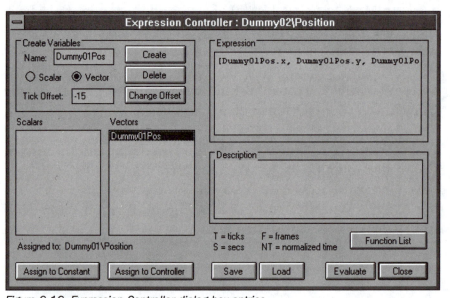

Figure 6-16. Expression Controller dialog box entries

Click Evaluate.

Create three more dummy objects, Dummy03, Dummy04, and Dummy05. Repeat Step 10 for each one, using the same expression to define the position of each dummy object. Substitute the following values for the Tick Offset.

| Object | Tick Offset |
|---|---|
| Dummy03 | -30 |
| Dummy04 | -60 |
| Dummy05 | -120 |

## Step 11. Create light beam clones

Select the light Direct01. From the Edit menu, choose Clone. Specify Instance as the Clone type.

Under the Motion Panel, click on Pick Target under the LookAt Parameters rollout. Choose Dummy02.

Repeat the above steps three more times. The table below shows the name of the dummy object each directional light should be looking at.

| Light Name | Looking at |
|---|---|
| Direct02 | Dummy02 |
| Direct03 | Dummy03 |
| Direct04 | Dummy04 |
| Direct05 | Dummy05 |

### Step 12. Set up copies as volume lights

After creating these new lights, they must be assigned as volume lights in order to show up in the rendering.

Under the Rendering menu, choose Environment. Under Atmosphere in the Environment dialog box, click Add. Then, in the Add Atmospheric Effect dialog box, choose Volume Light and click OK.

Under the Volume Light Parameters rollout, click Pick Light. Select Direct02, Direct03, Direct04, and Direct05.

Under the Rendering menu, choose Video Post. Choose Execute Sequence and click on the Render button.

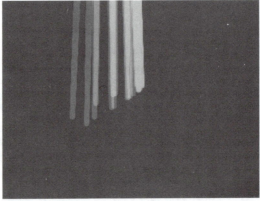

*Figure 6-17. Multiple laser beams*

## Step 13. Make shape visible

Now that our laser beam looks better, you'll notice that it is difficult to see the shape being traced. To remedy this, you'll create a separate object to represent the shape.

Create a copy of M-Spline. Under the Modify panel, click Edit Spline. Choose Spline as the sub-object level.

Under the Edit Spline rollout, turn on the Center checkbox and click on Outline. Type in 1.0 in the Outline Width field and press <Enter>.

Click on Sub-object. Click Edit Mesh.

Click Edit Stack.  In the Edit Modifier Stack dialog box, click Collapse All. Cick Yes in the Warning dialog box. Click OK to exit the Edit Modifier Stack dialog box.

*Adding an Edit Mesh modifier to the shape creates a surface.*

## Step 14. Material

Now you need a material to apply to this object. With the M shape still selected, click on the Material Editor button to open the Material Editor window. Create a material with a bright green Diffuse color and 100% Self-Illumination. Click on the

Assign Material to Selection button.

### Step 15. Opacity expression controller

Next you'll assign an Expression Controller to the Opacity track of the material. The expression is going to use the values of the Ease Curve you created earlier to control the rate at which the material's opacity changes.

Let's take a moment to work out what we want to have happen. Visually in the Track View, an Ease Curve varies between 0 and 100, but its actual numerical representation is that value multiplied by 160. This number comes from the fact that there are 160 ticks per frame. So at a value of 100, the Ease value is really reporting the value 160 x 100 = 16000 to the Expression Controller.

The Opacity track shows values between 0 and 100, but its actual values are fractions between 0 and 1. If the Opacity value in Track View shows 60, the number that is actually passed on to the Expression Controller, is 0.6, or (60/100).

Our Ease Curve varies from 0 to 6500, so we know through a simple proportion that our highest Ease Curve value must be (6500*160) or 1040000. Divide the Ease Curve value by 1040000 to bring the value down into the range that the opacity track will accept.

You may be wondering where the 0.6 comes into all of this. The expression is multiplied by 0.6 to decrease the intensity of the opacity. By doing this, not only are you guaranteed that the opacity level will never go above 60, but it decreases the slope of the curve as well.

We then subtract 0.2 from the expression to let the function curve go negative until about frame 195, which is the same time our laser beam really starts to pick up speed. If the Expression Controller passes a number to the Opacity track that's out of its valid range, it clips the value to 0 or 1, whether the value is above or below this range.

Doing this lets us automatically vary the intensity of its opacity to the ease curve, so as the laser beams move faster and faster, the material becomes more opaque.

Click on the Track View button ![icon] to open the Track View window. Expand the hierarchy tree so you can see all the parameters of the material.

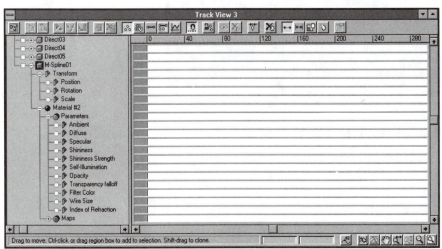

Figure 6-18. Material parameters hierarchy tree

Click on Opacity. Click the Assign Controller button. ![icon] Choose Float Expression from the list and click OK.

Right-click on the Opacity track. The Expression Controller dialog box appears. Create a Scalar variable. Name the variable Ease.

Click on the Assign Controller button 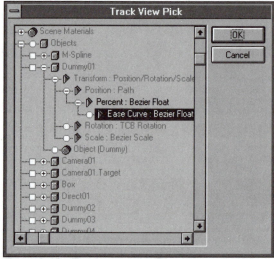 and highlight Ease Curve:Bezier Float as shown in Figure 6-19.

*Figure 6-19. Choosing the ease curve*

Click OK. Type the following into the Expression field.

***((Ease/1040000)\*0.6)-0.2***

Click Evaluate.

## Step 16. Render final animation

The laser beam now moves along the M shape, leaving a trail of semi-transparent beams as it goes along. The animation is ready to render.

Save your work as LASERSHO.MAX.

Under the Rendering menu, choose Video Post. Click on Add Image Output Event. Click on Files and enter the name LASERSHO.AVI. Click OK to exit the dialog box. Click Execute Sequence. Enter a Range of 0 to 299. Click Render to start rendering the animation.

## Things to Try

This example is just scratching the surface of what can be done with laser beams in MAX. I urge you to take the techniques presented here and explore them to fit your own animation needs. Try some shapes of your own, or try different color combinations. It's up to you!

# c h a p t e r

## Vehicle Over Terrain

In this chapter, you'll create a vehicle and roll it across bumpy terrain. The rotation and movement of the wheels and vehicle body will be automatically calculated by using many tools and tricks.

The finished rendering can be viewed in the files VEHCLE_1.TGA and VEHCLE_2.TGA in the \KYLE\IMAGES directory on the CDROM.

In Tutorials 16 and 17, you'll create an off-road vehicle and suspension system. Tutorial 18 shows you how to create terrain with displacement mapping. In Tutorials 19 and 20, you'll move the vehicle across the terrain along a path. With numerous expressions, every aspect of the motion will be calculated automatically.

*TIP* *To do all the steps in these tutorials, you will need Release 1.1 or above of 3D Studio MAX. To see which version of MAX you have, choose Help/About. A dialog box appears. The version number appears at the top of the dialog box. The version number Release 1 refers to Release 1.0.*

*Even if you don't have Release 1.1 or above, you can still do these tutorials. However, you'll have to skip a few steps and your model won't be as complete as the final renderings shown.*

*All registered users of Release 1.0 are entitled to a free upgrade to at least Release 1.1. See the Resources section of the Appendix under Kinetix for information on how to get an upgrade.*

Figure 7-1. Vehicle model

Although these tutorials are intended to be done in sequence, you can jump to Tutorials 17, 18 or 19 without doing previous tutorials. The models required for these tutorials are supplied on the CDROM.

## Tutorial 16
## Building an Off-Road Vehicle

This tutorial explores some techniques you can use to create a model of an off-road vehicle. The model created here will be used in later tutorials when the vehicle is animated to move across bumpy terrain.

The off-road vehicle consists of three sections: the body, the wheels, and the suspension system. In this tutorial you'll create the body and wheels. The suspension system is created in the next tutorial.

### Step 1. Preparation

Load the file VEH_MODl.MAX from the \KYLE\MESHES directory on the CDROM.

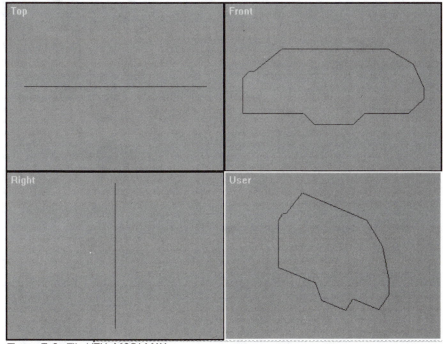

*Figure 7-2. File VEH_MODI.MAX*

This model, containing several dummy objects and splines, is provided to save you time in doing this tutorial. The only object displayed is a spline called BodySpline. You'll unhide the other objects as needed over the course of this tutorial.

### Step 2. Vehicle cab

Select BodySpline. Under the Modify Panel, choose Extrude. In the Extrude Parameters Rollout, set the Amount value to -40.0 and the Segments value to 3. Click Zoom Extents All.

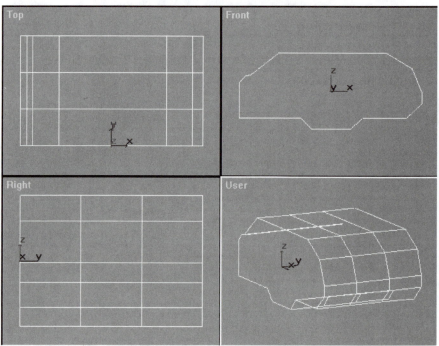

*Figure 7-3. Extruded BodySpline*

Click on Edit Mesh. The Vertex level is selected as the Sub-Object level.

Right-click in the Top viewport to activate it. Select the center two sets of vertices on the object as shown below.

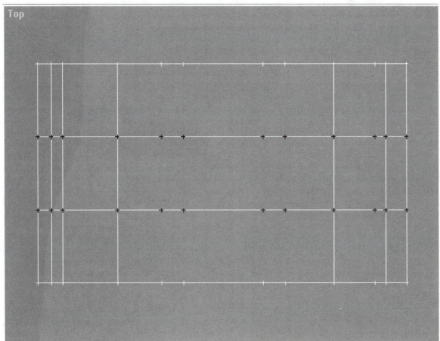

*Figure 7-4. Selected vertices*

Click Select and Non-uniform Scale  from the toolbar. Click

on the Lock Selection Set button to turn it on.

*When the selection set is locked, the Lock Selection Set button turns yellow. You can also lock the selection by pressing <spacebar> on your keyboard.*

Turn on Restrict to Y 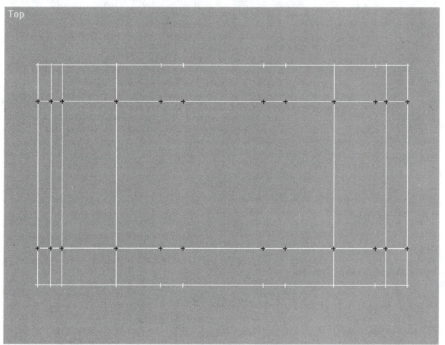 to affect only the Y axis. Turn on Percent Snap. In the Top viewport, click and drag to scale the vertices to 200%. Refer to the numbers on the status bar to make sure you scale the vertices to exactly 200%.

Figure 7-5. Scaled vertices

Press the spacebar to unlock the selection set. Deselect vertices so that only the front three sets of middle vertices are selected. Check the user viewport to make sure you have the right set of vertices selected.

*To deselect vertices, hold down the <Alt> key while drawing the selection region. A minus sign appears next to the cursor when deselection is active. You must hold down the <Alt> key during the entire time you draw the selection box.*

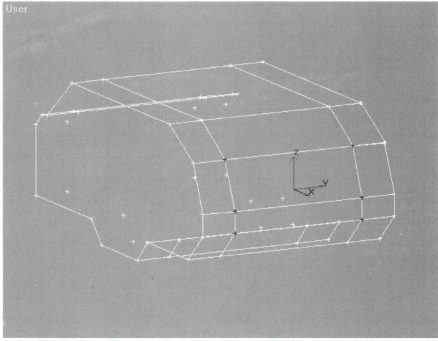

*Figure 7-6. Selection set*

Click Select and Move. 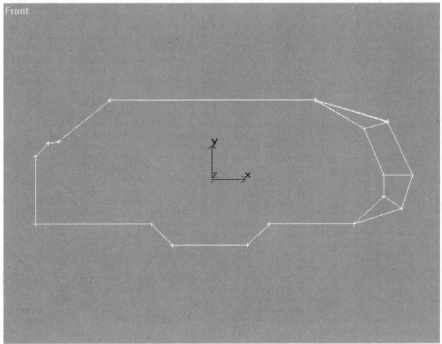 In the Front viewport, move the vertices to the right as a group, then move different groups to extend the windows of the cab area as shown below.

Figure 7-7. Front set of vertices moved

## Step 3. Windows

The windows are made by extruding one of the cab's faces.

Choose the Face level of Sub-Object. Under the Edit Face rollout, make sure Polygon is chosen as the selection type. Click on the center polygon of the cab to select it. The easiest way to do this is to click on the face in the User viewport.

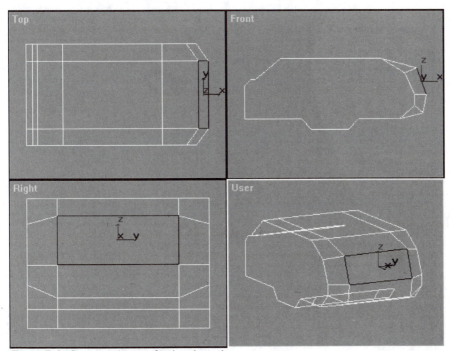

*Figure 7-8. Center polygon of cab selected*

Under the Edit Face rollout under the Extrusion section, enter 0.001 in the Amount field.

A very small number is used for this value so the window will stay virtually coplanar with the rest of the mesh.

*Material ID 2 was used for the windows because Material ID 1 will be taken by the material for the cab. Changing the face's Material ID will allow you to assign a material to the windows different from the cab material.*

Click Select and Uniform Scale. 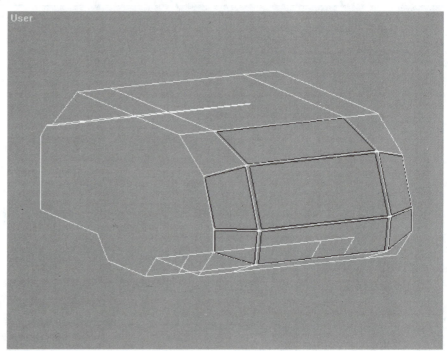 Turn off Percent Snap. In any viewport, scale the selected face to 95%. Under the Edit Surface rollout, set the Material ID to 2.

Repeat the above steps for the polygons above and to the side of the already scaled polygon. Also repeat the steps for the bottom row of polygons on the cab for a total of seven windows.

*Figure 7-9. Extruded cab window polygons*

Click Sub-Object to turn it off.

## Step 4. Mapping

Now you'll assign UVW mapping coordinates to the object in preparation for the Multi/Sub-Object material later on.

Under the Modify panel, click UVW Map. Specify the mapping type as Planar in the UVW Map parameters rollout. Click Sub-Object to activate the gizmo.

Right-click in the Front viewport and turn on the Angle Snap Toggle. Click Select and Rotate and turn on Restrict to Z. In the Front viewport, rotate the UVW Map gizmo by 180 degrees.

> *You can also activate angle snap by pressing the <A> key on your keyboard.*

Click Sub-Object to exit the gizmo sub-object level of the UVW Map modifier.

Click on Fit. Fit matches the extents of the mapping coordinates to the extents of the object to which they are applied.

## Step 5. Collapse mesh

Next you'll collapse the vehicle cab into one entity. All material ID assignments and mapping coordinates are retained when the mesh is collapsed.

> *Collapsing the mesh will free up memory and allow the screen to redraw faster.*

Click on Edit Stack. In the Edit Modifier Stack dialog box, click on Collapse All. Answer Yes to the warning. Click OK to exit the dialog box.

Rename BodySpline to VehicleBody.

### Step 6. Create panel

There are other pieces of the vehicle that can be added to make the vehicle more interesting. Some of it will be created with geometry, while other components will be suggested with the materials you apply to the vehicle.

These objects serve no practical function, but they give us an opportunity to explore three different modeling techniques: extrusions, the bevel modifier, and boolean operations.

Under the Display panel, click Unhide by Name. Choose Spline1 from the list.

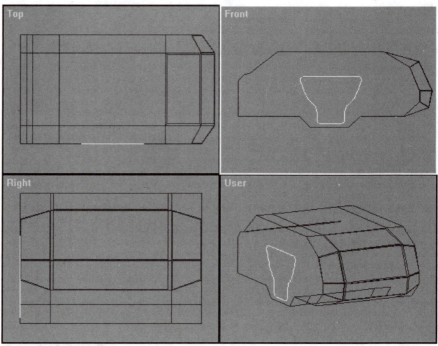

*Figure 7-10. Spline1*

A spline appears. This spline will used to make a raised panel.

Select Spline1. Under the Modify panel, click on More. Choose Bevel from the list. Under the Bevel Values rollout, enter the settings in the table below.

| | |
|---|---|
| **Start Outline** | 0 |
| **Height** | -0.1 |
| **Outline** | 0 |
| **Level 2** | ON |
| **Height** | -0.1 |
| **Outline** | 0.05 |
| **Level 3** | ON |
| **Height** | 0.02 |
| **Outline** | 0.05 |

You can see the result of the bevel in the Top and Right viewports.

Activate the Front viewport. Under the Modify Panel, click UVW Map. Click Sub-Object. Click Select and Rotate.

In the Front viewport, rotate the UVW Map gizmo by 180 degrees. Watch the status line at the bottom of the screen to see how far the gizmo is rotated.

Turn off the Sub-Object level. Collapse the object's stack.

### Step 7. Create another panel

Under the Display Panel, click Unhide by Name. Choose Spline2 from the list. This object is a set of closed splines that can be extruded all at once.

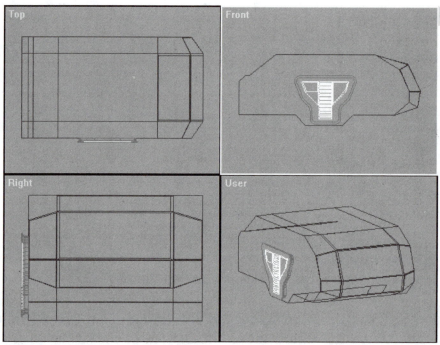

*Figure 7-11. Spline2*

Select Spline2. Under the Modify panel, click Extrude. Under the Parameters rollout, enter 0.05 for Amount and set the Segments to 1.

Click UVW Map. Right-click in the Front viewport. Click Sub-Object. Click Select and Rotate. In the Front viewport, rotate the UVW Map gizmo by 180 degrees.

Collapse the object's stack.

### Step 8. Create more panels

Let's add one more element. For this one, you'll subtract one extruded spline from another.

Under the Display panel, click Unhide by Name. Choose Spline3a and Spline3b from the list.

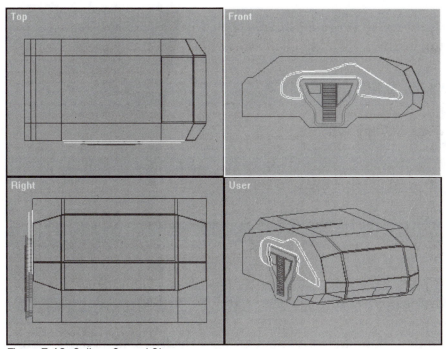

*Figure 7-12. Splines 3a and 3b*

Select Spline3a, the outer spline, and go to the Modify panel. Click Extrude. Set the Extrude Amount to -0.2.

Select Spline3b, the inner spline. Under the Modify panel, click Extrude. Set the Extrude Amount to -0.5.

Select Spline3a again. Under the Create panel, choose Geometry. Select Compound Objects from the pulldown list.

*When picking Spline3b for the boolean operation, you might find it difficult to pick the spline from the screen. To pick the operand, you can press the <H> key and pick from a list instead.*

Click Boolean on the Object Type rollout. On the Parameters rollout under the Operations section, make sure Subtraction (A-B) is selected.

Click Pick Operand B. Select Spline3b.

Under the Modify panel, click UVW Map. Click Sub-Object.

Right-click on the Front viewport. Click Select and Rotate. Rotate the UVW Map gizmo by 180 degrees in the Front viewport.

Collapse the object's stack.

## Step 9. Copy objects

Now that we've created some objects for one side of VehicleBody, we need to copy them to the other side.

Select Spline1, Spline2, and Spline3a. Activate the Top viewport.

Click Mirror Selected Objects. In the Create Mirror Object dialog box, set the Mirror Axis to Y. Set the Offset value to 42.140.

Under Clone Selection, click Copy. Click OK to exit the dialog box.

The vehicle body is now complete. Save your work as VEHC01.MAX.

## Step 10. Tire

Now that the body is completed, you need some wheels. First you'll create a tire. A torus primitive will suffice as a base object. The torus will then be modified to make treads on the tire.

Under the Create panel, click Geometry. Choose Standard Primitives from the pulldown list. Click on Torus. Create a torus in the Front viewport as shown below.

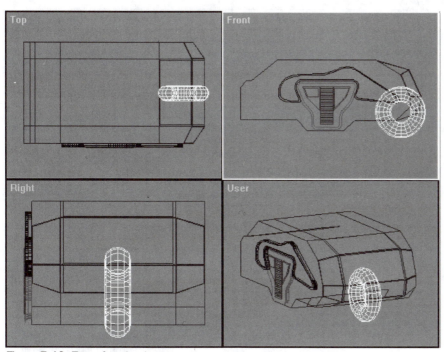

*Figure 7-13. Torus for wheel*

*The parameters here will make a very smooth tire. If you're short on CPU speed or RAM, you might want to cut the number of sides and segments in half. This will speed up redraw time when you ani-mate the vehicle later on. Be sure to use an even number of seg-ments so you can make the tire treads later in this tutorial.*

Enter the following parameters for the torus.

| | |
|---|---|
| **Radius 1** | 5 |
| **Radius 2** | 3 |
| **Segments** | 32 |
| **Sides** | 18 |

Name the torus Tire.

## Step 11. Treads

Now you'll edit the tire to reshape it and make treads.

Select the tire. Under the Display panel, click Hide Unselected. This will hide everything except Tire to make it easier to work

Click Zoom Extents All. Under the Modify panel, click Edit Mesh. In the Sub-Object Selection level, select Face. In the Top viewport, select one-half of the faces as shown in Figure 7-14.

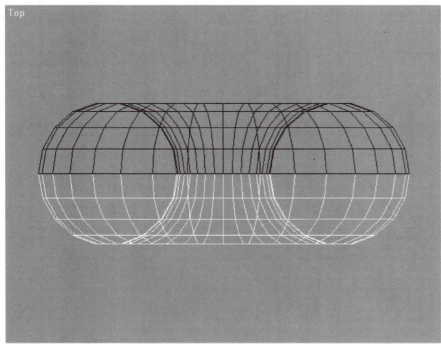

*Figure 7-14. Top half of faces selected*

Under the Edit Face rollout in the Extrusion section, set the Amount to 4.0. This shapes the torus more like a tire.

Next you'll delete some of the torus's interior faces. These faces won't be seen in the final rendering, and will only increase redraw and render time if left in the model.

Set the selection method in the toolbar to Circular Selection Region. 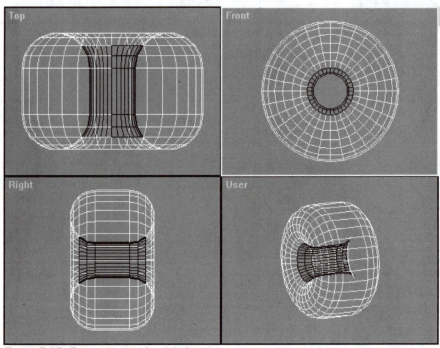 In the Front viewport, select faces as shown below.

*Figure 7-15. Face selection for deletion*

> 🎵 **You can also delete the faces by clicking Delete on the Edit Surface rollout under the Miscellaneous section.**

Press the <Delete> key on your keyboard to delete the faces. Click Yes to delete isolated vertices.

Set the selection method to Fence Selection Region. 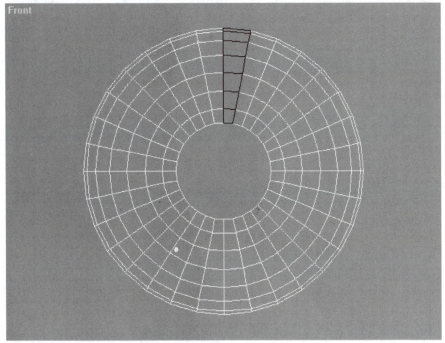 Click

the Window/Crossing Selection toggle ⬚ to turn on Window Selection if necessary.

In the Front viewport, draw the fence selection to select one strip of faces as shown below.

*When the <Ctrl> key is depressed just before selection, a small + appears next to the cursor. In order to add to your selection, you must make sure the + appears before you start selecting. You must also hold down the <Ctrl> key during the entire time you draw each fence. Otherwise, you won't add to your selection; you'll create a new selection, losing your previous one.*

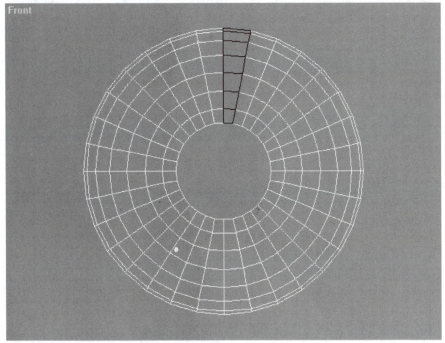

*Figure 7-16. One strip of selected faces*

Add to your selection to select every other strip of the tire, as shown in Figure 7-17. To do this, hold down the <Ctrl> key on the keyboard as you select each strip.

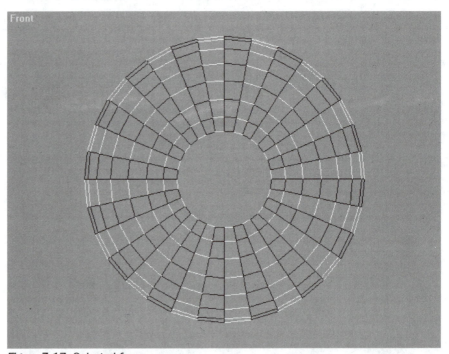

Figure 7-17. Selected faces

*Here, you have increased the radius of the wheel by 0.5 units. The total radius of the wheel is Radius 1 + Radius 2 + extrude value. This is 5.0 + 3.0 + 0.5 for a total of 8.5 units. This number will be very important to us when we animate the model in later tutorials.*

Now you'll extrude the selected faces to make treads.

On the Edit Face rollout under the Extrusion section, enter 0.5 for the Amount. The selected faces are extruded to make treads on the tire.

Click Sub-Object to turn it off.

## Step 12. Smooth tire

The tire is a little rough-looking. To smooth it out, you'll use a modifier called MeshSmooth.

Click the More button to bring up a list of additional modifiers. Choose MeshSmooth from the list. MeshSmooth takes a few moments to calculate the new surface.

In the Surface Parameters section of the Parameters rollout, turn on the Smooth Result checkbox.

By default, MeshSmooth is applied with one iteration. This setting provides adequate results for general use of this model. If you plan on rendering views close up to the tires, you may want to use an iteration of 2.

*MeshSmooth is available only in Release 1.1 or higher of 3D Studio MAX. If you don't have Release 1.1 or higher, see the Resources section of the Appendix, under Kinetix, to find out how to get it. In the meantime, you can't do this step of the tutorial, but you can continue on to the next step and complete the tutorial.*

*Meshsmooth adds faces to the mesh. If you use an iteration of 2 or higher, a significant number of faces will be added to your mesh. This slows redraw and rendering time.*

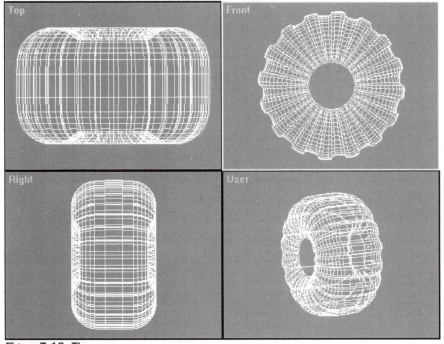

Figure 7-18. Tire

### Step 13. Finish tire

Now you'll set up the tire to receive a mapped material, and collapse the mesh to save memory. Under the Modify panel, click UVW Map.

Collapse the object's stack.

### Step 14. Hub

The tire now needs a hub.

Under the Create panel, click Shapes. Click Circle.

In the Front viewport, create a circle of radius 3.0. Position it flush with the tire as shown in Figure 7-19.

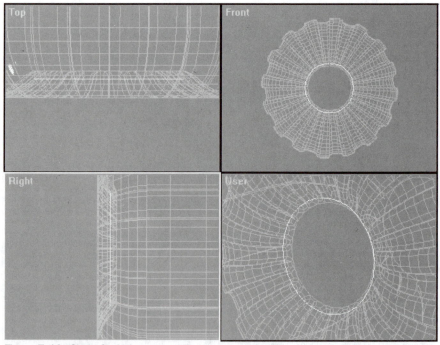

*Figure 7-19. Circle for hub*

Under the Modify panel, click More. Choose Bevel from the list. Set the following parameters for the Bevel modifier.

| | |
|---|---|
| **Start Outline** | **0.0** |
| **Height** | **0.2** |
| **Outline** | **0.2** |
| **Level 2** | **ON** |
| **Height** | **-0.1** |
| **Outline** | **-0.45** |
| **Level 3** | **ON** |
| **Height** | **-0.1** |
| **Outline** | **-0.1** |

Name the object Hub.

To check the positioning of the hub, change the User viewport to Smooth + Highlight. Move the hub if necessary so the hub is flush against the tire and the entire hub shows.

This object will also have a mapped material. Under the Modifier panel, click UVW Map.

Collapse the object's stack.

### Step 15. Copy hub

Next you'll create a copy of the hub for the opposite side of the wheel.

Activate the Right viewport and select Hub if necessary. Click Mirror Selected Objects. Set the Mirror Axis to X. Under the Clone section of the dialog box, click Copy. Set the offset value to 8.5.

Click OK to exit the dialog and create the copy.

### Step 16. Attach hubs to tire

Select Tire. Go to the Modify panel. Under the Edit Object rollout, click Attach Multiple. Choose Hub and Hub01 from the list. The hubs are now attached to the tire.

The wheel is now complete. Rename the tire/hubs object to WheelL1.

### Step 17. Wheel material

Even though each element of WheelL1 has its own mapping co-ordinates, it is not quite ready to receive materials yet.

Choose Face in the Sub-Object selection level.

Under the Edit Face rollout, locate the Selection section. Click the Element button. You can now select faces by element.

Select both Hub elements. The easiest way to do this is to select all faces, and then deselect the Tire element. Under the Edit Surface rollout, change the Material ID for these faces to 2. Turn off Sub-Object.

Since the tire and hubs have different Material IDs, you will be able to assign two different materials to them even though they are part of the same object.

> **TIP** *Material IDs can be assigned to faces either before or after you attach them to another object. Faces retain their Material IDs through all mesh modifications, including booleans.*

### Step 18. Pivot point

When you extruded the first selected set of faces, this caused the object's pivot point to move away from the center of the object. You'll need to put the pivot point back in order for the wheel to work properly.

With WheelL1 selected, go to the Hierarchy panel. Under the Adjust Pivot rollout, click Affect Object Only.

*By clicking on Affect Object Only, you'll retain the position of the object's local axis.*

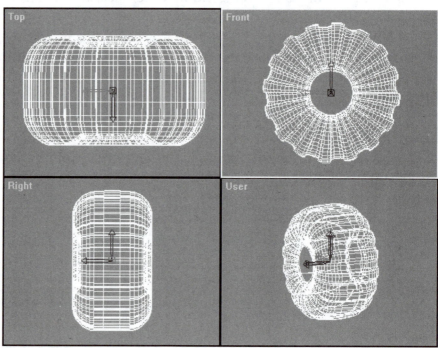

Figure 7-20. Pivot point

Click Center to Pivot. The object is now centered on the pivot.

The first tire is now complete. Save your work as VEHC02.MAX.

**Step 19. Wheel copies**

Now that we have one completed wheel, we need to make three copies for the other wheels on the left side. The position of each wheel needs to match the positions of its dummy object, so we'll need to unhide the dummy objects as well.

Under the Display Panel, click Unhide by Name. Select the following objects from the list.

**DumWheelL1**
**DumWheelL2**
**DumWheelL3**
**DumWheelL4**
**DumWheelR1**
**DumWheelR2**
**DumWheelR3**
**DumWheelR4**

Zoom Extents All. These objects are the dummy objects that will control each wheel's position.

Select WheelL1. Click on Align. Click Select by Name. Select DumWheelL1 from the list. Turn on X Position, Y Position, and Z Position. Choose Pivot Point for Current Object and Target Object. Click on OK to exit the Align Selection dialog box.

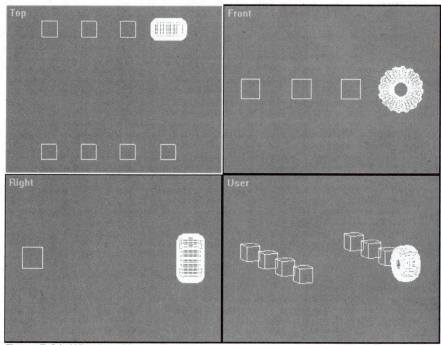

Figure 7-21. Wheel position

Create three copies of the wheel. Name the wheels WheelL2, WheelL3 and WheelL4. Use the same procedure to position the wheels with their corresponding dummy objects.

Create four more copies of the wheels with the names WheelR1, WheelR2, WheelR3 and WheelR4. Position them over their dummy objects in the same way.

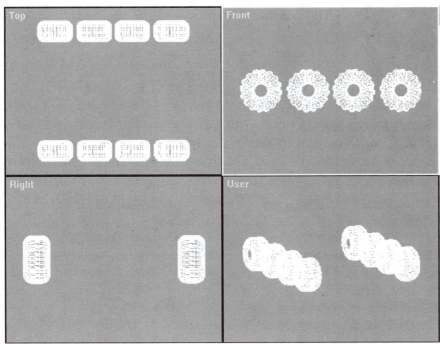

*Figure 7-22. Eight wheels in position*

Hide the wheel dummy objects to clean up the screen. Select all objects starting with DumWheel, and use Hide Selected to hide them.

Unhide VehicleBody, Spline1, Spline2 and Spline3a and Spline3a01 (the mirrored copy of Spline3a).

The vehicle is now complete. Save your work as VEHC03.MAX.

### Step 20. More stuff

Just one finishing touch and you'll be ready for a weekend get-away. Some extra objects have been supplied for you to strap onto the top of the vehicle.

These extra objects been created for you. All the objects were created with basic methods similar to the ones you've seen in this tutorial.

Under the Display panel, click Unhide by Name. Unhide the object Extra Stuff.

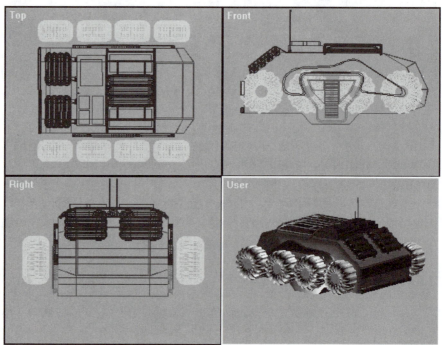

Figure 7-23. Extra stuff on vehicle

Save your work as VEHC03.MAX, overwriting the file you saved in the previous step.

The extra objects have all been collapsed so you can work with them as one object. These extra objects can be found in the file VEH_XTRA.MAX in the \KYLE\MESHES directory on the CDROM. The objects in VEH_XTRA.MAX are not collapsed, so their modifier stacks are still visible and editable.

If you're interested in a particular object, you can load VEH_XTRA.MAX to see how it was made. The majority of the objects are cylindrical or rectangular solids, or slight variations thereof, and are mapped with cylindrical, box, or planar mapping coordinates.

# Tutorial 17
# Modeling the Suspension System

In a later tutorial, the vehicle and wheels will be animated to follow a bumpy terrain. The vehicle motion will be offset from the wheel motion to simulate the way a real vehicle works. When a bump is encountered, the wheels move over it. A few frames later, the vehicle itself adjusts its height to compensate.

These delays in motion give us the opportunity to put a suspension system into the model. Eight simple helixes provide the springs. With one dummy object linked to      each wheel and another linked to the body, you can determine the top and bottom positions of each shock at any given time.

The actions of the springs can be set up with expression controllers. This tutorial sets up all the expression controllers for the eight springs on the vehicle. This setup will make it extremely easy to implement the suspension system later on. Once the vehicle is bumping along the terrain, letting the top dummy objects follow the vehicle, and bottom dummy objects follow the wheels, will put the suspension system in motion.

This tutorial refers frequently to "the animation". There is no animation in this tutorial, but in the next tutorial you'll be moving the vehicle and wheels around the terrain.

Some parts of this tutorial deal with expression controllers. It is recommended that you read Chapter 33 of the 3D Studio MAX User's Guide before going further.

To make this tutorial easier, take a few moments to reacquaint yourself with some of MAX's controller tools.

Load the file VEHC03.MAX from the previous tutorial. If you haven't done the previous tutorial, load the file VEH_MOD2.MAX from the \KYLE\MESHES directory on the CDROM.

Select any object. Under the Motion panel, expand the Assign Controller rollout. The transform parameters for the object's position, rotate and scale appear in a list.

Highlight one of the transform parameters. The Assign Controller button becomes available. You will be working with this rollout and button several times over the course of this tutorial.

Click Assign Controller to see the controllers available for this parameter. Click Cancel to exit the dialog box.

Next, click on the Track View button at the upper right of the screen. The Track View window appears.

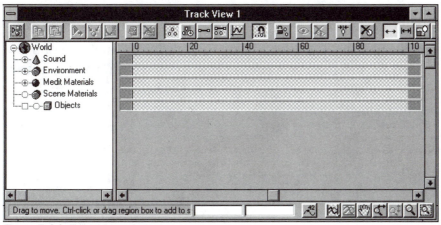

Figure 7-24. Track View window

This window shows all the animation activity for all parts of the scene.

*If you're working with a screen resolution of 800x600, you might have to pan the toolbar all the way over to the left to find the Track View button at the right end of the toolbar.*

Click on the + next to Objects. A list of all parent objects appears. To find more objects, locate an object with two + signs next to it. Click on the leftmost + to see their child objects.

Move the mouse in the Track View listing until the Pan icon appears. Click and drag to move the display down and to the left so you can see all the object names.

To see even more detail, locate an object with only one + sign and click on the +. Another level appears. One of the entries is called Transform. Click on the + next to Transform.

Here are the Position, Rotation and Scale tracks for the object. This level is where you'll be doing a lot of the work in this tutorial. The position and rotation of the wheels and vehicle body will be changed in a variety of ways right here in Track View.

Locate the Assign Controller button at the upper left of the Track View window. Also locate the Properties button at the upper right of the window. Again, you might have to pan the toolbar to find the Properties button as it is the rightmost button on the toolbar. These two buttons are used repeatedly throughout this tutorial.

Now that you know where your tools are, close Track View and begin.

## Step 1. Preparation

If you've done the previous tutorial,    load the file VEHC03.MAX.

If you haven't done the tutorial, reload the file VEH_MOD2.MAX from the \KYLE\MESHES directory on the CDROM.

In this tutorial you'll be working primarily with dummy objects for the wheels, so most of the model can be hidden for now.

Click Select by Name and select all objects except WheelL1 and VehicleBody. Hide the selected objects.

### Step 2. Create dummies

First you'll need two objects to control the suspension system.

Under the Create panel, click Helpers, then Dummy. Click and drag in the Front viewport to create a Dummy object as shown in Figure 7-25. The dummy object should be kept small in size so you don't clutter up your scene.

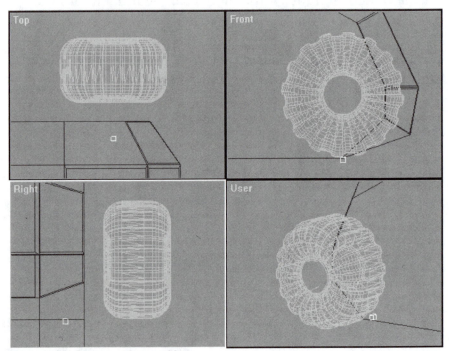

Figure 7-25. Dummy object position

Name the object DumLoL1. Note the L1 notation on the end of the object name. This denotes that the object is part of the first shock assembly on the left side of the vehicle.

Under the Edit menu, choose Clone. In the Clone Options dialog box, select Copy. Enter the name DumHiL1.

Position DumHiL1 and DumLoL1 near the front left wheel, as shown below.

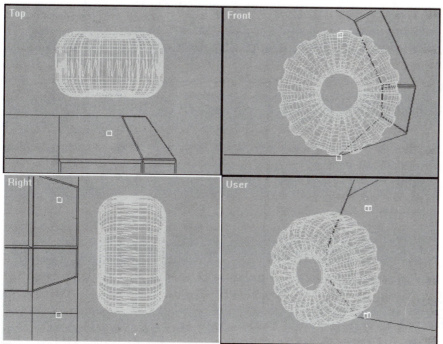

*Figure 7-26. Dummy positions*

Now that you have the dummies positioned, you no longer need to see the wheel. Hide WheelL1 and VehicleBody. You should now have just the two dummy objects on your screen.

### Step 3. Spring

As you might suspect, the suspension system will incorporate a spring. A helix primitive will work well for this purpose.

Under the Create Panel, click Shapes, then Helix. Create a Helix in the Top viewport.

Edit the helix parameters as shown in the following table.

| | |
|---|---|
| **Radius 1** | **2.50** |
| **Radius 2** | **2.0** |
| **Height** | **-12.0** |
| **Turns** | **8.0** |

Name the Helix SpringSplineL1.

For now, you'll leave the helix as is. Later on, you'll loft a circle along the helix to make a solid spring.

*The controllers you'll assign to the helix object can only be assigned to it because the helix is a parametric 3D spline. This means properties such as height are controlled by a parameter of the object and not as a function of scale. You can't apply these expression controllers to a 3D mesh object because a 3D object is not parametric and can only be controlled through generalized functions such as position and scale.*

### Step 4. Expression for spring

As noted before, this suspension system will work in such a way that you can control everything through the motions of the two dummy objects. To make this work, you'll need to relate various aspects of the other objects to the positions of the dummy objects.

SpringSplineL1's position must be made to follow DumHiL1's position. You might think you can accomplish this with linking, but in this case, expression controllers are necessary because if the dummy object's positions are changed via a link, its positions in world space are not accurately passed on to other expression controllers. If we have the helix match the position of the dummy object via an expression controller, the coordinate positions are correctly passed on to the expression controller.

In preparation for this acrobatic feat, you'll use an expression controller to make SpringSplineL1 follow DumHiL1. Later you'll find out how to make the other end of the spring follow the other dummy object.

An expression controller uses an equation, or expression, to control the movement of an object. The expression can utilize motion from one object to define the motion of another. Expression controllers work with variables which are set up as you go along. A variable can be set up to use any aspect of any object's animation.

Make sure SpringSplineL1 is still selected. Under the Motion panel, expand the Assign Controller rollout.

Under the Assign Controller rollout, highlight Position: Bezier Position.

Click the Assign Controller button on the Assign Controller rollout. The Replace Position Controller dialog box appears.

*Replacing Bezier Position with Position Expression means you can now use an expression to control the position of the object over the course of the animation.*

**TIP** *Throughout this tutorial a naming convention is used for variables in expression controllers. The variable name will be determined by taking the name of the controller that it is going to call and adding a notation that tells us what type of track it is calling. For example, the variable DumHiL1Pos denotes the Position track for DumHiL1.*

Choose Position Expression from the list.

In the Assign Controller rollout, click, then right-click on Position: Position Expression. A selection list appears. Choose Properties from the list. The Expression Controller dialog appears.

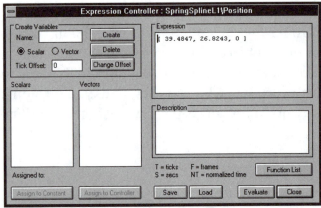

*Figure 7-27. Expression controller dialog box*

The numbers displayed in the Expression window on your screen may not match the numbers shown in Figure 7-27. The numbers in your window reflect the current position of your helix wherever you drew it on the screen.

**TIP** *If you type in one of the variables incorrectly, the Expression Parse Error dialog box appears when you click Evaluate. If this happens, check your entries for errors. Make sure the variable name DumHiL1Pos was entered correctly, and check that the expression you entered uses the variable name correctly.*

In the Create Variables section of the dialog box, type in DumHiPos. Set the Variable type as a Vector. Click Create. The variable name shows up in the Vectors list.

Typing directly in the Expression box at the upper right of the dialog box, edit the current expression to read:

**[DumHiL1Pos.x, DumHiL1Pos.y, DumHiL1Pos.z]**

Click Evaluate. MAX calculates for a moment or two, then the cursor appears again.

A Position Expression notates position in the form of [x, y, z] (refering to the locations on the x, y, and z axes) in the controller. We can use only one of three coordinate axes (x, y, or z) for each part. A vector variable carries these three components in it already. We can access these components individually by using a ".x", ".y", or ".z" notation after the vector variable name.

Click Assign to Controller at the bottom of the Expression Controller dialog box. A Track View Pick list appears. Expand the Objects listing and locate DumHiL1 on the list. Expand DumHiL1 and then expand Transform. Select the Position: Bezier Position track and click OK to exit the dialog box.

Click Close to exit the dialog box.

Now, wherever you move DumHiL1, SpringSplineL1 will move as well.

*Don't confuse Assign to Controller at the bottom of the dialog box with the Assign Controller button on the Motion panel and Track View window. The Assign Controller button lets you choose a type of controller for a track. Once you've used the Assign Controller button to set up an expression controller, such as Position Expression, you then set up the expression. Assign to Controller on the Expression Controller dialog box assigns a variable to a controller.*

### Step 5. Spring orientation

We want SpringSplineL1 to lock its other end to DumLoL1. But how can we lock that end to DumLoL1, when its position is already locked to DumHiL1?

There are two problems that need to be solved in order to make this relationship work properly. First, the orientation of SpringSplineL1 must be controlled to make it always point from one dummy to another. In addition, the length of SpringSplineL1 must be able to change over the course of the animation to fit right between DumHiL1 and DumLoL1.

*A Look At controller takes over the entire Transform controller of an object, and forces that object to change its rotation so that it constantly looks at another object.*

First, let's solve the orientation problem with a Look At controller. This controller, when assigned to SpringSplineL1, will cause the spring to "look at" DumLoL1 no matter where it is. Since one end of the spring is locked to DumHiL1, this controller will cause the spring to point directly from DumHiL1 to DumLoL1.

With SpringSplineL1 selected, go to the Motion Panel. Under the Assign Controller rollout, highlight Transform: Position/Rotation/Scale. Click the Assign Controller button  on the rollout. Choose Look At from the list.

On the Look At Parameters rollout, click Pick Target. Click on DumLoL1.

Now you can freely move both DumHiL1 and DumLoL1 around. SpringSplineL1 will always be oriented between the two dummies.

## Step 6. Spring length

You can also use an expression controller to calculate the height of the spring at any time in the animation. An expression can find the distance between the two dummy objects and pass that value on to the height parameter of the helix primitive. Wherever the two dummy objects move, SpringSplineL1 will always have its height determined by the distance between the two dummy objects. In this way, each end of the spring appears to be anchored at one or the other dummy object.

Click on Track View. The Track View window appears. Expand the Objects listing. Expand SpringSplineL1. Expand Object (Helix) under SpringSplineL1. Highlight the Height track.

Click the Assign Controller button on the Track View window. Select Float Expression from the list.

Highlight Height: Float Expression, then right-click on it. A selection list appears. Choose Properties from the list to bring up the Expression Controller dialog box. You are now going to define an expression for the height of the spring.

*If your Track view shows just Height and not Height: Float Expression, click Filters and turn on the Controller Types checkbox in the Show section.*

Create a Vector variable named DumLoL1Pos. The variable name appears under the Vectors list. Highlight the DumLoL1Pos variable name on the list and click Assign to Controller. Expand the object DumLoL1 and then expand Transform. Choose the Position: Bezier Position track of DumLoL1 from the list and click OK. This gives the position of DumLoL1 to the variable DumLoL1Pos at all times during the animation.

Create another Vector variable DumHiL1Pos. In the same way, assign this variable to the Position: Bezier Position track of DumHiL1.

Now that you have a variable for the position of each dummy object, you'll use those in an expression for the spring.

Enter the following expression in the Expression area:

*-(length(DumHiL1Pos - DumLoL1Pos))*

Click Evaluate, then Close.

The *length* function takes the length of a vector. If the length function is given a vector operation such as adding or subtracting vectors, it will return the length of the new vector. By subtracting one dummy position vector from another, the expression defines a new vector between the two dummy objects. This length can then be assigned to the spring height. The height will change over the animation to correspond to the changing distances between the two dummy objects.

Now, wherever you move the two dummy objects, SpringSplineL1 will reorient and change height accordingly. Feel free to move the dummy objects around to test the expressions.

*Take care to enter the length function with a lower case L. A capital letter at the beginning of length will cause the expression to fail.*

## Step 7. Create hydraulic components

Next you'll create hydraulic components to move up and down with the spring. Two cylinders will act as housing for the spring. These objects will move similarly to SpringSplineL1.

Create a cylinder with the following values:

| | |
|---|---|
| **Radius** | **2.0** |
| **Height** | **-6.0** |
| **Height Segments** | **1** |
| **Cap Segments** | **1** |
| **Sides** | **12** |

Name the cylinder CylL1a.

Create another cylinder with the following parameters:

| | |
|---|---|
| **Radius** | **1.50** |
| **Height** | **-12.0** |
| **Height Segments** | **1** |
| **Cap Segments** | **1** |
| **Sides** | **10** |

Name the cylinder CylL1b.

### Step 8. Assign cylinder positions

Select CylL1a. Under the Motion panel, expand the Assign Controller rollout. Highlight Position: Bezier Position and click the Assign Controller button. Choose Position Expression from the list.

Highlight and right-click on Position: Position Expression and choose Properties from the list. The Expression controller dialog box appears.

Create a Vector variable named DumHiL1Pos. Make sure you click Create to make the variable appear on the Vectors list.

Highlight the variable on the Vectors list, and click Assign to Controller. Choose the Position track of DumHiL1 and click OK.

Edit the expression to read:

*DumHiL1Pos*

Click Evaluate, then click Close. CylL1a's position is now locked to the position of DumHiL1.

Select CylL1b. Choose Position: Bezier Position from the list and click the Assign Controller button. Choose Position Expression from the list.

Highlight and right-click on Position: Position Expression and choose Properties from the list. The Expression controller dialog box appears.

Create a Vector variable named DumHiL1Pos. Highlight the variable on the Vectors list, and click Assign to Controller. Choose the Position track of DumHiL1.

Edit the Expression to read:

**DumHiL1Pos**

Click Evaluate, then click Close.

## Step 9. Assign Look At controllers

Next you'll assign a Look At controller to each cylinder to make them look at DumLoL1.

Select CylL1a. Under the Motion panel, highlight Transform:

Position/Rotation/Scale. Click the Assign Controller button. Choose Look At from the list and click OK.

Click Pick Target in the Look At Parameters rollout. Select the object DumLoL1.

Select CylL1b. Do the same operation to assign the same Look At controller to CylL1b.

### Step 10. Scale hydraulic component

Now the top cylinder of our hydraulic structure is working like SpringSplineL1 with the exception that its height is not scaling. In this step, you'll scale one of the cylinders to fit between the dummy objects.

Click Track View [icon] to display the Track View window. Expand the hierarchy to locate the Height track of CylL1b.

Highlight the Height track and click the Assign Controller button. [icon] Choose Float Expression from the list.

A black bar appears next to several items, including the Height track. Right click on the black bar next to the Height track. The Expression Controller dialog appears.

Create two Vector variables, DumHiL1Pos and DumLoL1Pos.

Select DumHiL1Pos from the Vectors list and click Assign to Controller. Choose the Position: Bezier Position track of DumHiL1 and click OK.

Select DumLoL1Pos from the Vectors list and click Assign to Controller. Choose the Position: Bezier Position track of DumLoL1 and click OK.

Edit the Expression to read:

*-(length(DumHiL1Pos- DumLoL1Pos))*

Click Evaluate, then click Close.

### Step 11. Limit spring and cylinder height

When the vehicle animation is complete, a situation might occur where the dummy objects move so much that the distance between them is less than the length of the fixed-height cylinder. This would cause parts of the assembly to look very strange.

To avoid this problem, the changes to the height of the spring can be limited.

In a previous step we created an expression controller for the length of the SpringSplineL1:

**-(length(DumLoL1Pos-DumHiL1Pos))**

This expression returns the distance between the two dummy objects. Something needs to be incorporated into the expression to test the distance between the two dummy objects and compare it to the height of the fixed-length cylinder.

If the distance is greater than the height of the fixed-length cylinder, the length of the spring should be the distance between the two dummy objects. If the distance is less than the height of the fixed-length cylinder, the value should be the height of the fixed-length cylinder.

The *if* function can be used to test the result of an expression. The *if(c,t,f)* function is a conditional function that tests if an expression is true or false, and returns a value accordingly. If c is true, the expression returns the value t. If c is false, the function returns f.

The *if* expression will be used in this way:

**[if (distance between dummies is less than cylinder height, use cylinder height, use distance between dummies)]**

Before continuing, look at this expression and the description of *if(c,t,f)* and make sure you understand the relationship between the two.

With this in mind, the Height expression for SpringSplineL1 can be changed to:

**if(-(length(DumLoL1Pos - DumHiL1Pos)) < CylL1aHeight, CylL1aHeight, -(length (DumLoL1Pos - DumHiL1Pos)))**

You should also take into consideration that you don't want the spring to ever become quite as short as the fixed-length cylinder. You'll want to leave a little clearance by limiting the height of the spring to the height of the cylinder plus a small amount, such as 1.5 units.

In this case, the expression becomes:

**if(-(length(DumLoL1Pos - DumHiL1Pos)) < CylL1aHeight + 1.5, CylL1aHeight + 1.5, -(length (DumLoL1Pos - DumHiL1Pos)))**

Next you'll implement the *if* statement into our previous expression controllers for SpringSplineL1 and CylL1b. First you'll need to set up CylL1a's Height track so we can call it from the expression controller.

In order for you to call another track from an expression controller, that track needs to have a controller assigned to it. You can create a key to cause a controller to be assigned to the track. Let's create a key for CylL1a's Height track.

Select CylL1a. Go into Track View. Click the Filters button.

Under the Show Only section, turn on the Selected Objects checkbox. Expand the hierarchy to show the parameters of CylL1b.

Click the Add Keys button. Click across from Height to create a key. Exit Track View.

Now you can enter the new expression for SpringSplineL1.

Go into Track View. Expand the hierarchy to show the Height track for SpringSplineL1.

Right-click on the range bar across from Height: Float Expression, bringing up the Expression Controller dialog.

Add a Vector variable named CylL1aHeight. Click Assign to Controller. Choose CylL1a's Height track.

Edit the expression to read:

*if(-(length(DumLoL1Pos - DumHiL1Pos)) < CylL1aHeight + 1.5, CylL1aHeight + 1.5, -(length (DumLoL1Pos - DumHiL1Pos)))*

Click Evaluate, then click Close.

Perform these steps for CylL1b as well.

### Step 12. Loft spring

Even though you've defined SpringSplineL1's motion with several animation controllers, it's not yet ready to render. SpringSplineL1 is still a spline, which has no surface. You can use the spline as the path of a loft object.

Go into the Create panel, and click Shapes. Click on Circle, and create a circle in any viewport with a radius of 0.2. Name the circle SpringShape.

Select SpringSplineL1. Under the Create panel, click Geometry. Choose Loft Object from the pulldown list.

Click Loft. Under the Creation Method rollout, turn on Instance. Click Get Shape, then click on the circle SpringShape.

Under the Skin Parameters rollout, turn on Skin. Set Shape Steps to 1 and Path Steps to 3.

Name the object SpringL1.

Note that you now have two spring objects, the un-renderable SpringSplineL1 and your newly created loft object SpringL1.

SpringL1 doesn't have the Look At Controller or the Position Expression Controller    that SpringSplineL1 does. The controller must be assigned again to SpringL1.

Go into Track View and expand the hierarchy to access the position track of SpringSplineL1. Select Transform and click Copy. Pan down in the hierarchy and expand SpringL1 so you can access its Position track. Select the Transform track and click Paste. Choose Instance in the dialog box and click OK.

Now the settings for SpringL1 match all the settings for SpringSplineL1.

If you want to change the radius of the coil, you need only edit the radius of SpringShape.

The suspension system is now complete. Save your work as VEHC04.MAX.

## Step 13. Seven more shocks

You can create seven more shock assemblies in the same way the first one was created. Simply substitute L2, L3, L4, R1, R2, R3, and R4 in place of L1 for each object or variable name.

I realize this would take a while, so I've provided the other seven assemblies in a file on the CDROM. To bring them into your model, merge all objects from the file SHOCKS.MAX from the \KYLE\MESHES directory on the CDROM.

Don't forget to save the model again once you've merged the shock assemblies. Hide all the dummy objects and save your work as VEHC05.MAX.

## Step 14. Materials

Materials for all parts of the vehicle have been provided on the CDROM.

I used Adobe Photoshop and Fractal Design Painter to create the maps. The source maps are included on the CDROM in the directory \KYLE\MAPS.

Click on Material Editor.  The Material Editor window appears. Click on Get Material. Under the Browse From section, choose Material Library. Click Open. Choose the library VEHICLE.MAT from the \KYLE\MAPS directory of the CDROM.

The following table shows the object names and the materials that should be assigned to them.

| Object Name | Material Name |
|---|---|
| VehicleBody | VehicleBody |
| WheelXX | Wheel |
| Spline1 | Spline1 |
| Spline2 | Spline2 |
| Spline3 | Spline3 |
| CylXXa | Gold Metal |
| CylXXb | Shock Metal 1 |
| SpringXX | Spring |
| Terrain | terrain |

Follow these steps for each material listed in the table.

Click on Get Material. Choose the material from the list. Select the corresponding object in the scene. Click Assign to Selection.

Save your work as VEHC05.MAX, overwriting the file you saved in the last step.

Render a perspective view of the vehicle to see the materials. Feel free to explore any of the materials with the Material Editor to see how each material was created.

There it is, the off-road vehicle. Next you'll create a bumpy terrain for it to roll along.

# Tutorial 18
# Creating Terrain with Displacement Mapping

In this tutorial you'll create a terrain mesh from scratch, and will use this mesh for the vehicle animation in later tutorials.

It's also possible to use a previously created terrain model for later tutorials, but a displacement map must first be generated from the terrain. The procedure for this technique is described in the section after this tutorial, *Using an Existing Terrain Mesh*.

## Step 1. Create grid

Under the Create panel, click Geometry. Choose Box and expand the Keyboard Entry rollout. Enter the following values in the Keyboard Entry rollout.

| | |
|---|---|
| X | 500 |
| Y | 500 |
| Z | 0 |
| Length | 1000 |
| Width | 1000 |
| Height | -500 |

Click the Create button under the entries.

Under the Parameters rollout, change the detail level of the box as follows.

| | |
|---|---|
| Length Segments | 100 |
| Width Segments | 100 |
| Height Segments | 1 |

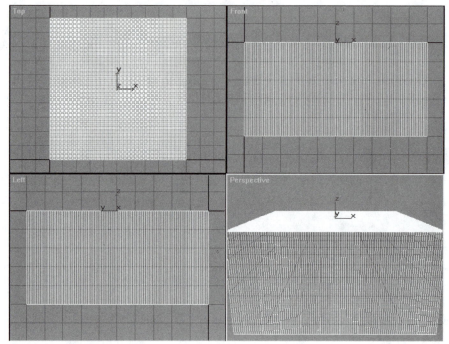

*Figure 7-28. Terrain grid*

Rename the box to Terrain. Click Zoom Extents All.

In a real life project, you might want to use a more detailed box. However, as you increase the number of segments, the software's performance slows down.

## Step 2. Remove bottom vertices

You need only the top surface of the grid, so the bottom and side faces can be removed for better performance. The easiest way to do this is to delete the vertices at the bottom of the grid.

Click on Zoom Extents All. Click on Sub-Object. Select the bottom set of vertices on the box by dragging a selection quad around them.

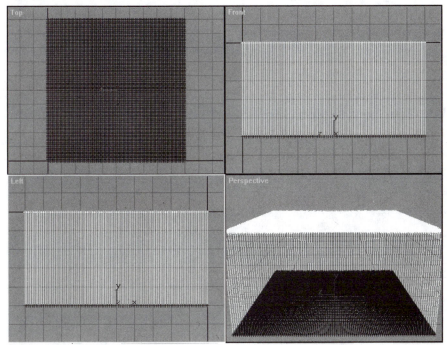

*Figure 7-29. Selected bottom vertices*

Click the Delete button under the Miscellaneous rollout. Click Sub-Object to turn it off.

Collapse the object's stack.

## Step 3. Displacement map

A displacement modifier uses a bitmap image to change the surface of an object. For example, a bitmap can be created with white areas to represent mountaintops, black areas for valleys and gray in the areas in between. When an image like this is applied to a flat object, such as our terrain, the mesh distorts to correspond to the bitmap.

Here you'll use a grayscale image supplied on the CDROM to distort the terrain grid. To view the image, choose View File from the File menu. Choose the file TERRAIN.TGA from the \KYLE\MAPS directory on the CDROM.

To use this bitmap to displace the terrain grid, click the More button under the Modifiers rollout at the top of the panel. Choose Displace from the list. Under the Parameters rollout, click the None button under Image. Choose TERRAIN.TGA from the \KYLE\MAPS directory on the CD ROM.

## Step 4. Set Strength

The Strength value is very important when assigning displacement maps. The Strength value defines the number of units of displacement that correspond to pure white on the bitmap. Pure black corresponds to zero units of displacement. All gray values in between correspond to various numbers of units, depending on the value of the gray.

For this bitmap, set the Strength to 150. When you're using an existing terrain model, there are calculations you must make to determine the appropriate strength value. This technique is described in the next section.

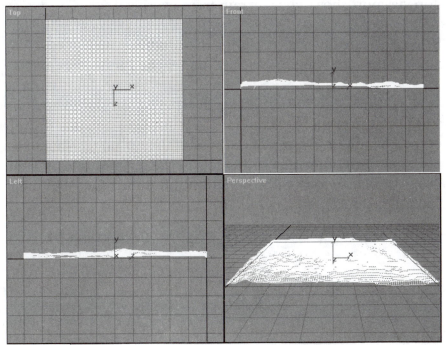

Figure 7-30. Displaced mesh

The terrain model is now complete. Save your work as TERR01.MAX.

## Using an Existing Terrain Mesh

Presented here is a method that you can use to extract height data from an existing terrain mesh.

The information needed from your terrain model is a height or displacement map. To extract this data from the terrain, you can project a gradient bitmap through the mesh and render a orthographic view. The gradient is white at the top and black at the bottom.

To create a displacement map from your terrain, load your terrain into 3D Studio MAX. These instructions assume that the Top viewport shows your terrain from overhead.

Select the terrain object. Under the Modify panel, click UVW map. Make sure Planar is selected under the Parameters rollout.

Turn on Sub-Object. Align the gizmo perpendicular to any one side of your terrain by rotating the gizmo. Make sure that the top of the gizmo is at the top of the mesh. The gizmo tab should be pointing in the direction of the top of the terrain mesh. Fit the gizmo to the terrain with the Fit button under the Alignment section of the Parameters rollout.

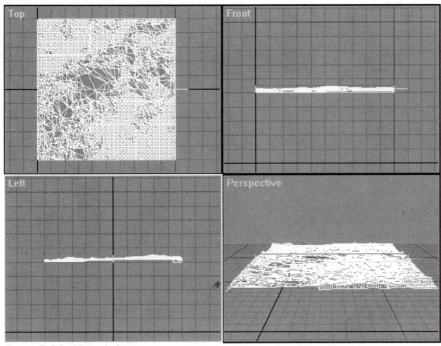

Figure 7-31. Aligned gizmo

Click on Material Editor.  Choose a sample box that is not being currently used.

Click on the Maps rollout. Click on the empty button next to Diffuse, and choose Bitmap from the Material Map Browser.

In the Bitmap rollout, click on the file selection button next to Bitmap. From the CDROM, choose the file V_GRAD.TGA from the \KYLE\MAPS directory.

Click Go To Parent. Set the Shininess Strength value to 0. Set the Self Illumination to 100.

**TIP** *The rendering from the Top viewport will have some black areas around the displacement map. These edges will have to be cropped off with a paint program before the image can be effectively used as a displacement map.*

Click Assign Material to Selection. Exit the Material Editor.

Activate the Top viewport. Click Zoom Extents. Render the Top viewport. Higher areas are more white, while lower areas are darker.

Render the Top view to a file. This image will have to contain adequate information from your terrain. When you render a top view of your mesh, you should have at least as many pixels in your image as you have grids in your object. Unless your terrain is very large, 500x500 pixels in resolution should work fine for rendering image size.

With this rendered image, not only could you recreate your terrain mesh, but you can displace an object's path so it will conform to the terrain.

Now that you have a displacement map of your existing mesh, you need one more piece of information: its height. This is the only way you'll know what strength value to use with the Displace modifier.

Go to either the Left or Front viewport. Press <Shift-B> to activate Box Mode. The display now shows a box for each object on the screen. You'll measure the height of the terrain with a tape helper.

Under the Create panel, click Helpers. Click Tape.

Click and drag the tape measure from the bottom to the top of the terrain. Note the length in the rollout.

Jot down the length somewhere you can find it later. I often use the text area in the Summary Info dialog to record notes about my scene. You can access the Summary Info dialog by choosing File/Summary Info from the drop-down menu.

Delete any merged objects such as the vehicle body and path.

If you want to use this terrain mesh with the vehicle animation, the grid will have to be moved to the appropriate location. Use your best judgement on where to position your terrain. You'll want to pay close attention so that you don't have the vehicle driving over areas of terrain that would be impossible for it to go over. If your terrain is in the neighborhood of 1000x1000 units, you'll want to position it at X:500, Y:500 so it will lie in the same general area as the terrain created in the previous tutorial.

*The vehicle model's scale is one unit = four inches.*

Save your terrain mesh.

## Tutorial 19
## Vehicle Motion Across Terrain

Interaction between objects in a computer generated scene can be the most influential aspect of the entire animation. When one object acts upon another, how does it act, or react? Bad simulation and interaction between multiple objects can destroy the effect of the scene. Likewise, correct modeling of interactions between objects can give your animation the edge it needs to be convincing.

In this tutorial you'll learn how to create and animate an off-road vehicle model as it drives over uneven terrain. We'll take into account many different aspects of our vehicle's motion, such as surface detection between the wheels and the terrain, calculations for the wheel rotations, the movement and rotation of the vehicle's body based on the positions of the wheels, and the motion of the suspension system.

### Step 1. Preparation

Before you start working with MAX, install a plug-in utility. From the CDROM, copy the file SPLENGTH.DLU from the \KYLE\PLUGINS directory to your 3DSMAX\PLUGINS directory. This plug-in will be used later on in this tutorial.

If you've done Tutorial 17, load the file VEHC05.MAX.

If you haven't done Tutorial 17, you can load the model from the CDROM. The file, called VEH_MOD3.MAX, can be found in the \KYLE\MESHES directory of the CDROM.

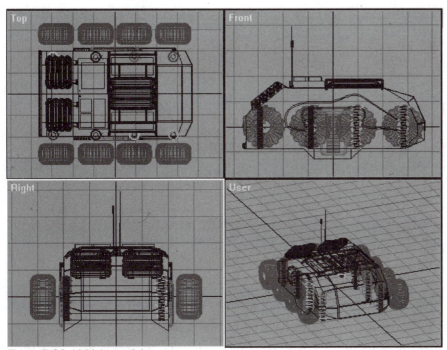

*Figure 7-32. Vehicle model*

This file contains the vehicle body, wheels, suspension system, all necessary dummy objects, and a path for the vehicle to move along. It also has all the expression controllers for the suspension system built in.

### Step 2. Merge terrain

The terrain must be merged into the file so the displacement information can be used to build the movement expressions.

If you did Tutorial 18, merge the Terrain object from the file TERR01.MAX. If you haven't done the tutorial, you can use a terrain mesh supplied on the CDROM. Merge the Terrain object from the file \VEH_TERR.MAX from the \KYLE\MESHES directory on the CDROM.

The vehicle is positioned at the lower left corner of the terrain as viewed from the Top viewport.

### Step 3. Hide objects

The terrain mesh is necessary in this tutorial for extracting certain measurements, but it's not necessary to see it all the time. You can hide the terrain to make your work go faster. You can also hide the wheels for the first part of this tutorial.

Select the terrain grid and the wheels. Under the Display panel, click the Hide Selected button under the Hide by Selection rollout.

### Step 4. Make dummy follow path

The total number of frames must be set to 300 for this animation. To do this, click on the Time Configuration button  and set the End Time value to 300.

Our first step in animating this vehicle is to make the dummy object, DumBody, follow the path PathInitial.

Select DumBody. Under the Motion Panel, expand the Assign Controller rollout. Choose Position:BezierPosition.

Click on the Assign Controller button. Choose Path from the list. In the Path Parameters rollout, click on Pick Path and choose the path object PathInitial.

In the Path Parameters rollout, under Path Options, turn on Follow.

Move the frame slider back and forth. DumBody and VehicleBody are moving along PathInitial over the course of the animation.

### Step 5. Access position track

Next you'll affect the vehicle's vertical position, or ground clearance, as it moves along the path. Because the vertical position changes throughout the animation, you'll use a variable to calculate the vehicle's position at each spot on the terrain.

The Track View area of MAX can be used to modify an object's position according to a variable.

Click the Track View button. Click the + next to Objects. Note that some of the objects listed have two + signs next to them, while some have one.

The listings with two + signs are parent objects. Clicking the left + expands the list of children, while the right + leads to transforms and other controllers for the object itself.

The object VehicleBody is a child of DumBody, so click the left + next to Dumbody. Scroll down to find VehicleBody, and click the + sign on the right. Click the + next to Transform. The Position track for VehicleBody is now displayed.

Even though VehicleBody moves in the scene, there are no keys displayed for its position track. VehicleBody moves in the scene because it's linked to DumBody, which is moved along the path. Subsequently, DumBody currently has position keys but VehicleBody does not.

## Step 6. Assign Position List

The vertical position information for VehicleBody will be put on its own position track. For its position you'll use a Position List. A Position List lets you assign multiple animation controllers to one animation track, compounding the effects of each controller into a final value that is passed on to the position track. Since you'll be assigning more information to the vehicle body's position later on, you'll need a Position List so you can continue to add the information.

Before continuing, you'll expand the track names so you can always see which controller type you're using.

Click the Filters button ![icon] at the top left of the Track View window, and turn on the Controller Types checkbox. Next, expand the hierarchy to VehicleBody's Transform tracks.

Click on the name Position: Bezier Position to highlight it. Click the Assign Controller button. ![icon] Choose Position List from the list.

The name Position: Bezier Position has changed to Position: Position List, and a + sign has appeared next to it. Click on the + sign next to Position: Position List to expand the list. Bezier Position now appears on the list, along with another item called Available. This label is merely a marker for the next available slot in the Position List, making it easy for you to choose it later on if necessary.

### Step 7. Set up Position Expression

The Bezier Position entry won't be used for this tutorial. Instead you'll use a Position Expression. A Position Expression is an equation used to determine a value.

Highlight the Bezier Position entry. Click on the Assign Controller button. Choose Position Expression from the list. A series of lines appear next to VehicleBody, Transform, Position and Position Expression. Unlike a Bezier Position controller, an Expression controller does not have any specific keys, but rather is calculated on each frame based on the equation used in the expression controller. For this reason, its animation is represented by a range bar rather than keys.

### Step 8. Set up expression controller

> *You can also access the Expression Controller dialog box by right-clicking on the black range bar across from the track name.*

Highlight Position Expression and click the Properties button at the upper right corner of the Track View window. The Expression Controller dialog box appears.

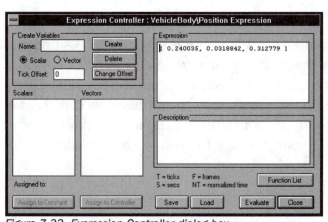

*Figure 7-33. Expression Controller dialog box*

Next you're going to write an expression to evaluate VehicleBody's vertical position. The first thing you need to do is set up a variable.

There are two types of variables in MAX's expression controller system; Scalars and Vectors. As you may recall from high school geometry class, a Scalar is simply a number, such as 22 or 911.74. A Vector has two components, a direction and a length. In MAX, a Vector is defined by (x,y,z) coordinates. It is assumed that the origin point is the (0,0,0) coordinate. From this information, MAX can glean the direction and length of the Vector variable.

Both types of variables can be assigned either a constant value or changing value. If either is assigned a constant value, its value remains unchanged over the length of your animation. If assigned to an animation controller, such as the position track of an object, the variable can change throughout the animation based on that object's position.

A valid Scalar could be a constant such as 422.71, or it could be assigned to a controller in your scene that returns one value for each frame of the animation, such as the radius or number of segments in a cylinder. If the radius or number of segments change over the course of the animation, the Scalar variable changes with it.

For your animation, you'll be using both Scalar and Vector variables. First you'll use a Scalar.

A Scalar variable called GroundClearance will be used to specify how high you want VehicleBody to ride above the ground. Right now you're not concerned with positioning the car with respect to the height of the terrain or the motion of the wheels. You simply want to position the car above the ground. Later on you'll change this variable to take the terrain into account. For now, you'll simply set up this variable to make the vehicle ride above the ground.

Click in the Name field and type in the word GroundClearance. Make sure Scalar is selected. Click Create, then click Assign to Constant. Enter 25 as the value and click OK.

Now that we have our variable set up, we need to write an expression to tell VehicleBody to ride the value of GroundClearance (in units) above the path.

Typing directly in the Expression box at the upper right of the dialog box, edit the current expression to read:

**[0, 0, GroundClearance]**

Click Close. Note that VehicleBody has changed vertical position.

Save your work as VEHC06.MAX. Go on to Tutorial 20 to set up the motion of the wheels.

# Tutorial 20
# Wheel and Suspension Motion

In this tutorial you'll set up the wheel motion and rotation. Many different aspects of the vehicle's motion, will be taken into account, such as surface detection between the wheels and the terrain, calculations for the wheel rotations, the movement and rotation of the vehicle's body based on the positions of the wheels, and the movement of the suspension system.

This tutorial picks up where Tutorial 19 left off. Load the file VEHC06.MAX if it is not already on your screen.

## Step 1. Creating paths for wheels

Unhide the eight wheel dummy objects, all objects beginning with DumWheel. Move the time slider to see how the dummy objects move with the vehicle body.

Each of the eight wheel dummy objects are moving correctly across the ground plane in X and Y space, but the Z axis value is unchanging. This means the wheels are not moving up and down in relationship to the terrain.

The height of the terrain is different at the contact points of each of the eight wheels, so each wheel's Z axis motion must be treated separately. This means you need a separate spline path for each wheel so you can add the Z axis motion to each one independently.

Right now the wheels' motion in X,Y space is determined by the motion of the dummy object DumBody. You must generate an individual path for each wheel based on its actual motion before you can work with the independent Z axis values.

To generate the path for DumWheelL1, select the object DumWheelL1. Under the Motion panel, click on Trajectories. Make sure your Start and End values read 0 and 300 respectively. You should have at least half as many samples as you have frames, so for this animation, set Samples to 150. One sample per frame is actually better, but dealing with spline objects with very large numbers of vertices can slow performance.

Click Convert To. A new spline appears on the screen representing the wheel's path. The new spline is called Shape01. Select this new spline and change the name to PathWheelL1. It's very important to rename the spline so you can find it later. You'll be making splines for all eight wheels, and the default names will be too confusing.

Repeat this process for the remaining seven wheels, renaming the new spline path each time with the appropriate name as shown in the table below.

| Dummy Object | Path Name |
|---|---|
| DumWheelL1 | PathWheelL1 |
| DumWheelL2 | PathWheelL2 |
| DumWheelL3 | PathWheelL3 |
| DumWheelL4 | PathWheelL4 |
| DumWheelR1 | PathWheelR1 |
| DumWheelR2 | PathWheelR2 |
| DumWheelR3 | PathWheelR3 |
| DumWheelR4 | PathWheelR4 |

The reason for generating so many steps is so that when we displace all the paths, we have enough detail in the resulting path to correctly follow the ground (which is also being created by the same displacement map). A lesser amount of steps would result in the paths, and eventually, the wheels, intersecting the terrain.

Once we have the paths for all the wheels created, there is no reason for them to be linked to DumBody anymore.

Select all eight wheel dummy objects and click the Unlink Selection button.

## Step 2 Make wheels follow paths

Note that even though you created a path from each wheel dummy, the dummies don't actually follow the paths right now. When MAX generates a path from an object, it doesn't automatically make the object follow the path. Once you've unlinked the wheel dummy objects, move the frame slider back and forth. The wheel dummies now don't follow any paths at all. Each wheel dummy must now be made to follow its own path.

Select the wheel dummy object DumWheelL1. Under the Motion panel, expand the Assign Controller rollout. Highlight Position:Bezier Position and click the Assign Controller button.  Choose Path from the list.

Click the Pick Path button. Select the path PathWheelL1. Click OK. Turn on the Follow checkbox. The object DumWheelL1 now moves along its path.

Repeat this process for each wheel dummy object, referring to the table above for the corresponding path names.

*If you're using your own ter-rain mesh, set the Strength value to the height of the ter-rain in units.*

## Step 3 Displace paths vertically

Next you'll displace each wheel path so it matches the vertical height of the terrain as the wheel moves along.

If the terrain object is hidden, it must be unhidden for this step. Under the Display panel choose Unhide by Name, and choose Terrain from the list. The terrain mesh appears.

Select each PathWheel spline object. Under the Modify panel, click the More button. Choose Displace from the list.

Under the Alignment section near the bottom of the rollout, click Acquire. Click on the Terrain object. A small dialog box appears. Choose Acquire Absolute and click OK.

The gizmo now matches the size and orientation of the terrain. This ensures that when the path is displaced, it will correctly match the terrain.

Click on the bitmap file selector button and choose TERRAIN.TGA from the \KYLE\MAPS directory on the CDROM. Set the Strength value to 150.

Note that the Strength setting matches the Strength used to create the mesh from the displacement map.

The paths are now displaced along the Z axis in accordance with the height of the terrain at each spot along the path. Move the animation slider and watch the Front viewport to see the wheel dummies move along the terrain in X, Y and Z space.

Now you'll collapse the modifiers for these objects. Click Edit Stack, ▤ then click Collapse All. Click Yes in the resulting dialog, and click OK.

Save your work as VEHC06.MAX.

## Step 4. Adding wheel objects

Now that our wheel dummy objects are moving correctly over the terrain, we need to add the wheels to our simulation.

Unhide all objects with names beginning with Wheel.

You'll notice that the X and Y positions of the wheels match the positions of the wheel dummy objects. Each wheel has the same position as its parent dummy object, and each of those dummy objects are following the exact topography of the terrain.

You can hide the terrain to save redraw time if you like.

### Step 5. Simplify tires

In this tutorial, many animation controllers are applied directly to the wheels themselves. This means the wheels must remain in the scene. Unfortunately, there is no other way to do this; the controllers won't work correctly if applied to substitute objects, dummy objects or instances.

You might find that the display takes an inordinately long time to redraw with the eight tires onscreen. If this is the case, there are a few changes you can make to improve performance.

One is to apply an Optimize modifier to each tire. When the animation is complete, you can then delete the modifier from each tire's stack to restore the original tires.

You can also hide all the wheels, then unhide them one at a time as needed. The disadvantage to this is that you won't be able to see all the animation effects while you're working.

The suggested method is to apply an Optimize modifier to each tire. You can select all the tires and apply the Optimize modifier to all tires at once. This will give you the greatest improvement in speed with the least amount of work. Try Optimize with a Face Thresh value of 12.

## Step 6. Reposition wheels

With their current paths, half of each wheel is underground. You'll need to use an expression controller to relate the vertical position of each wheel to the position of its dummy object so that each wheel will always ride on top of the terrain topology. This is a similar technique we used to relate the position of VehicleBody to the position of DumBody.

Each wheel dummy object's position in Z space is exactly the height of the terrain. If each wheel is going to ride the correct height above the terrain, you need to add the radius (or one-half the diameter) of the wheel to the Z position of its dummy.

First you'll set up an expression controller for this relationship.

Select the object WheelL1. Under the Motion panel, expand the Assign Controller rollout. Choose Position:Bezier Position from the list and click Assign controller button. Choose Position Expression from the list.

Highlight Position Expression, then right-click on Position Expression. Choose Properties. The Expression Controller dialog box appears. Click in the Name field and type the label Radius. Make sure Scalar is selected. Click Create.

You have just created a Scalar variable called Radius. Recall that each tire has a radius of 8.5 units.

Click Assign to Constant. Enter 8.5 as the value and click OK. Type the following into the expression controller:

*[0, 0, Radius]*

Because each wheel has the same relationship to its parent object, we can copy the controller we just created and copy it down to all the other wheel position tracks. The familiar Windows Copy and Paste functions can be used to copy the controller to the other wheels' position tracks.

Make sure WheelL1's position track is highlighted, and click the Copy button ▣ at the top of the Track View window.

Expand each of the other wheel's hierarchies so their position tracks are exposed. Highlight each wheel's position track by holding down the <Ctrl> key and clicking on each one.

Click the Paste button. ▣ Specify Instance in the dialog box and click OK.

## Step 7. Rotation equation

Now that the wheels are tracking the ground, they need to turn. If the vehicle is moving at a uniform speed throughout the animation, you can simply figure out how far it moves from beginning to end, and divide by the circumference of the wheel. This will give you the total number of full rotations each wheel makes over the entire animation.

In this case, you need to set up an expression to figure out the appropriate rotation rate at any given time during the animation. This expression uses the distance traveled on the path to determine the number of rotations, and rotates the wheel accordingly.

To express the wheel's rotation value at any frame during the animation, you can use the following expression:

*degToRad(-360 \*((PathLength \* PathPercent)/(2 \* pi \* Radius)))*

This equation might look confusing, but when you break it down into parts, it all makes sense. Let's take a few moments to look at the equation.

You can determine how far each wheel has moved on any given frame by taking the length of the path and multiplying it by the percent of that length the wheel has covered. This gives us a single, numerical value. Note that the PathPercent value is always changing during the animation.

*PathLength \* PathPercent = Distance Traveled*

343

The PathLength can be found with a utility included on the CDROM. PathPercent is a subordinate track of Path, so this number is always available to you. With these two values you can figure out how far the wheel has traveled at any point in the animation.

You also need to know how much to rotate the wheel on any given frame. For this, you can use the relationship between the distance traveled and the wheel's circumference. We already know that the radius of each tire is 8.5 units.

*(2 \* pi \* Radius) = Wheel Circumference*

MAX has a built-in variable called *pi* that already holds the pi value. You can use this variable in any expression as is.

You can divide the distance the wheel has traveled by the wheel's circumference to find out how many times it has rotated.

*(PathLength \* PathPercent)/(2 \* pi \* Radius) = number of rotations*

Then you can multiply the number of rotations by 360 degrees so you know the cumulative rotation. In this case, the wheel is rotating in the negative direction, so the number of rotations is multiplied by -360.

*-360 \* (PathLength \* PathPercent)/(2 \* pi \* Radius) = number of degrees rotated*

> **TIP** *Pi is a number roughly equal to 3.1416. If you snoozed your way through algebra and geometry in high school, you're going to have a hard time understanding this tutorial. If you're really lost, get a few math textbooks from the library and brush up your geometry.*

The *degToRad* function converts the value, in degrees, into radians for the rotation controller. Our final expression looks like this:

**degToRad(-360 \*((PathLength \* PathPercent)/(2 \* pi \* Radius)))**

This is the equation for figuring out the wheel's rotation on its local Z axis at any moment during the animation. So you can see that you have everything you need to set up this expression.

## Step 8. Spline length

First you'll use the plug-in from the CDROM to get the length of the path. Perform these steps for each wheel.

Select one of the path splines for a wheel. Under Utilities, choose Spline Length from the pulldown list.

Click Pick Spline. Click on one of the spline paths. The length of the spline then appears in the length field of the Spline Length Utility.

Jot the number down on a piece of paper, along with the name of the path.

Repeat this step for each wheel path spline.

*The Spline Length plug-in supplied on the CDROM does not work if you have modified the path by scaling it. Any other modifications performed on the path, such as modifiers, are taken into account.*

### Step 9. Set up variables

In order to express the rotation of the wheel along the Z axis only, you'll use another type of controller called a Euler XYZ controller. Euler, like Boole of boolean fame, is another mathematician whose name has crept into computer animation terminology. A Euler XYZ controller allows you to control each of the X, Y and Z axes of rotation separately.

The Euler XYZ controller will be applied to the Xform gizmo of the wheel.

Repeat the following steps for each of the wheels.

*Take care to choose Xform and not Linked Xform from the list.*

Select the wheel. Under the Modify panel, click More. Choose Xform from the list.

Open Track View. Expand the hierarchy tree several levels so you can see the Rotation: TCB Rotation controller of the Xform Gizmo. Highlight Rotation: TCB Rotation.

Click Assign Controller. Choose Euler XYZ from the list. The track name changes to Rotation: Euler XYZ.

Expand the Euler XYZ controller and highlight the Z Rotation.

Click Assign controller. Choose Float Expression from the list.

Highlight Float Expression and click the Properties button at the upper right corner of the Track View window. The Expression Controller dialog box appears.

For your convenience, the rotation expression has been included on the CDROM. Click Load. Choose WHLROTAT.XPR from the \KYLE\EXPRESSN directory on the CDROM.

Assign the following variables to their respective controllers.

| Variable | Type | Controller / Constant |
|---|---|---|
| PathLength | Scalar | Constant: Path length value returned by the Spline Length utility |
| PathPercent | Scalar | Controller: Path Percent track of the wheel's dummy (parent) object. |
| Radius | Scalar | Constant: 8.5 |

Now you have all the necessary data to make the wheels rotate properly.

Click Evaluate, then click Close.

Be sure to repeat this step for each wheel.

### Step 10. Vehicle ground clearance

When you move the frame slider back and forth, you'll see that the object VehicleBody is presently traveling underground. The expression you wrote for VehicleBody's ground clearance worked, but only when the ground was flat. Now that the terrain is bumpy and irregular, you need to go back and add something else to its expression controller that will allow it to ride a comfortable distance above the terrain at all points in the animation.

But how can you estimate this height? The wheels on each side are almost always at different heights. The best way is to average the heights of all the wheels and use this height for VehicleBody.

To make the animation realistic, it would be even better to make it look as if VehicleBody is reacting to the changes in height in the way a suspension system does. This can be done by setting the height of VehicleBody to the average height the wheels had a little earlier. All this can be done with expression controllers.

To average the heights, you'll need to set up variables for the wheel heights and use those to calculate the average. Remember the variable GroundClearance you created earlier? This variable will be replaced with an equation which averages the heights of the wheels a few frames earlier than the current frame.

To work with the timing, you'll use a time unit called a *tick*. In MAX, a tick is 1/4800 of a second. This fine division of time allows you to precisely control actions that take place in MAX.

For the purpose of this tutorial, we'll assume you plan to play back your animation at 30 frames per second (fps). In this way, we can work with ticks in a way that's easier to understand.

$$\frac{4800 \text{ ticks}}{30 \text{ frames}} = 160 \text{ ticks per frame}$$

With a playback speed of 30 fps, there are 160 ticks per frame. We want to estimate the height of the wheels 1.5 frames earlier, so we'll ask for the height of the wheels 240 ticks earlier.

Now you're ready to set up the variables for wheel height. Select VehicleBody. Under the Motion panel, expand the Assign Controller rollout. Choose Position:Position Expression. Right click and choose Properties.

Create the following new variables in the expression controller.

| Variable | Type | Assigned to Controller | Tick Offset |
|----------|------|------------------------|-------------|
| L1Pos | Vector | DumWheelL1:Position | -240 |
| L2Pos | Vector | DumWheelL2:Position | -240 |
| L3Pos | Vector | DumWheelL3:Position | -240 |
| L4Pos | Vector | DumWheelL4:Position | -240 |
| R1Pos | Vector | DumWheelR1:Position | -240 |
| R2Pos | Vector | DumWheelR2:Position | -240 |
| R3Pos | Vector | DumWheelR3:Position | -240 |
| R4Pos | Vector | DumWheelR4:Position | -240 |
| Radius | Scalar | Constant value: 8.5 | -240 |

With all these variable set up, you can now change the variable GroundClearance to a new equation averaging the heights of the wheels 1.5 frames ago.

Since the wheel position variables are vectors, they have three values as part of their definitions: X, Y and Z. You want to use only the Z position in your equation.

When you want to use just the Z value of a vector, express it as follows:

Vectorname.z

The average height of all wheels is the height of all wheels added together, then divided by eight. With this in mind, change the existing expression to read:

*[0, 0, ((L1Pos.z + L2Pos.z + L3Pos.z + L4Pos.z + R1Pos.z + R2Pos.z + R3Pos.z + R4Pos.z)/8) + GroundClearance + Radius]*

Click on Close.

The equation above adds the heights of all the wheels, divides them by eight, and adds the original GroundClearance and Radius variables to give the final height of the vehicle.

Move the frames slider back and forth to watch the vehicle body as it moves across the terrain. It follows the wheels closely but with a slight delay.

## Step 11. Adding noise

Now that VehicleBody is positioned correctly over the wheels, a little random motion, or *noise*, can be added to make the motion more realistic.

Select VehicleBody. Under the Motion panel, expand the Assign Controller rollout. Choose Available from the controller list and click the Assign Controller button. 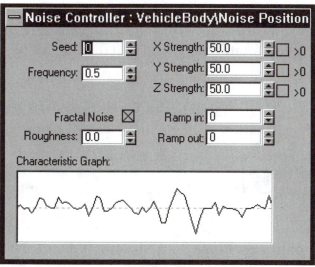 Choose Noise Position from the list.

Highlight Noise Position, then right-click on Noise Position and select Properties. The Noise Position dialog box appears.

Change the X and Y values to 0. Change the Z value to 1.50. Change Frequency to 0.15. Turn off the Fractal Noise checkbox.

> **When adding noise, take care not to set the values too high for the size of the object and the range of motion, or you may find your objects careening offscreen.**

Figure 7-34. Noise Controller dialog box

Close the dialog box. Move the frame slider back and forth. The vehicle body's height varies a little at random times.

### Step 12. Vehicle rotation

The vehicle body now moves up and down over the terrain satisfactorily. To make the animation more realistic, you can make VehicleBody rotate back and forth and from side to side in accordance with the wheel movements.

Over the course of the animation, VehicleBody might rotate along any of the two axes (x and y) depending on what's going on with the wheels. For this reason, you'll need a Euler XYZ controller for each of the two axes.

Open Track View. Locate VehicleBody, and expand it if necessary to display the Transform entry, then expand Transform to see the Rotation entry. Highlight Rotation. Click Assign Controller.

 Choose Euler XYZ from the list.

For the X axis rotation, you'll write an expression that takes the average height of the left side wheels and the average height of the right side wheels and rotates the gizmo based on the difference between the two.

Expand the Rotation listing. Highlight the X axis rotation track of VehicleBody's Xform gizmo.

Click Assign Controller. Choose Float Expression from the list.

Right-click on the controllers range bar to bring up the expression controller dialog box.

Set up the following variables.

| Name | Type | Controller | Tick Offset |
|------|------|------------|-------------|
| L1Pos | Vector | WheelL1:Position | -240 |
| L2Pos | Vector | WheelL2:Position | -240 |
| L3Pos | Vector | WheelL3:Position | -240 |
| L4Pos | Vector | WheelL4:Position | -240 |
| R1Pos | Vector | WheelR1:Position | -240 |
| R2Pos | Vector | WheelR2:Position | -240 |
| R3Pos | Vector | WheelR3:Position | -240 |
| R4Pos | Vector | WheelR4:Position | -240 |

Enter the following expression:

*(((L1Pos.z + L2Pos.z + L3Pos.z + L4Pos.z)/4)-(( R1Pos.z + R2Pos.z + R3Pos.z + R4Pos.z)/4)) / 50*

Click Evaluate.

Let's take a minute to break down what we're doing.

The first part of the expression,

*(L1Pos.z + L2Pos.z + L3Pos.z + L4Pos.z)/4*

is adding the heights (value on the Z axis) of the left side wheels and dividing it by 4 (there are 4 nodes). Then, we're doing the same thing on the right side:

*(R1Pos.z + R2Pos.z + R3Pos.z + R4Pos.z)/4*

We then find the difference between the two by subtracting the right side value from the left side value.

*((L1Pos.z + L2Pos.z + L3Pos.z + L4Pos.z)/4)-(( R1Pos.z + R2Pos.z + R3Pos.z + R4Pos.z)/4)*

The number we get, if passed directly on to the rotation controller will be very big, and cause VehicleBody to over-rotate. So, we dampen the effect by dividing everything by 50.

*(((L1Pos.z + L2Pos.z + L3Pos.z + L4Pos.z)/4) - (( R1Pos.z + R2Pos.z + R3Pos.z + R4Pos.z)/4)) / 50*

For the Y axis rotation, we'll use a similar approach, using the difference between the front and rear set of wheels.

In Track View, select the Y axis rotation track of VehicleBody.

Click Assign Controller. Choose Float Expression from the list.

Right-click on the controllers range bar to bring up the expression controller dialog. Set up the following variables.

| Name | Type | Controller | Tick Offset |
|------|------|------------|-------------|
| L1Pos | Vector | WheelL1:Position | -240 |
| L4Pos | Vector | WheelL4:Position | -240 |
| R1Pos | Vector | WheelR1:Position | -240 |
| R4Pos | Vector | WheelR4:Position | -240 |

Enter the following expression:

$$((( L4Pos.z + R4Pos.z)/2) - ((L1Pos.z + R1Pos.z +)/2))/50$$

Click Evaluate, then click Close.

This expression is almost identical to the Y rotation expression except that we're taking into account only 4 of the eight wheels, the frontmost and backmost sets.

## Step 13. Xform modifiers for wheels

Assign another Xform modifier to each wheel. Assign a Euler XYZ controller to the rotation track of each new Xform Modifier gizmo.

Assign a Float Expression controller to the X axis rotation track of WheelL1's second Xform modifier. Create a Scalar variable named VehicleBodyRot. Click Assign to Controller. ▶ Choose the X rotation track of VehicleBody and click OK. Edit your expression to read:

*VehicleBodyRot*

Click Close.

Go into Track View. Select the Rotation controller of WheelL1's second Xform modifier. Click Copy. Select each of the other Wheel's second Xform modifier's rotation track and click Paste. Choose Instance in the dialog box and click OK.

Save your work as VEHC07.MAX.

## Step 14. Suspension system

Next you'll unhide the suspension system. Under the Display panel, choose Unhide by Name. Select the objects listed below.

| | |
|---|---|
| DumHiL1 | DumLoL1 |
| DumHiL2 | DumLoL2 |
| DumHiL3 | DumLoL3 |
| DumHiL4 | DumLoL4 |
| DumHiR1 | DumLoR1 |
| DumHiR2 | DumLoR2 |
| DumHiR3 | DumLoR3 |
| DumHiR4 | DumLoR4 |

Each shock assembly is built so that all motion can be determined by the animation of two dummy objects. There are already expression controllers applied to each shock assembly to let you do this.

In a real vehicle, the suspension system plays a very large role in the way the chassis reacts to the topology of the terrain. The chassis of our vehicle is not influenced by a suspension system, but we simulated the influence with tick offset values in the expression controllers. This added latency to the motions of our chassis, simulating the delay of the force acted on the wheel as it travels through the suspension to the vehicle body.

We need to add an actual suspension system model to give the viewer the feeling that the suspension system is doing what it seems to. That's right, we're cheating.

### Step 15. **Generate suspension dummy paths**

In order to make the suspension system work, you'll need spline trajectories for the eight wheel objects and the eight dummy objects linked to VehicleBody. These splines will be used as paths for the two suspension dummy objects that control the motion of the spring and cylinders.

The lower shocks will follow paths from the wheels. The upper shocks will follow paths generated from eight dummy objects that sit under the vehicle body. These dummy objects' sole purpose is to generate paths for the upper shocks.

Select the dummy object that you want to extract the trajectory from. Under the Motion panel, click on Trajectories.

Make sure your Start and End values read 0 and 300, respectively. Set the number of samples to 150.

Click Convert To. Select the new spline by pressing the <H> key and selecting Shape01 from the list.

Change the Name of the spline object from Shape01 to the corresponding name in the table below.

*If you rename each new spline object as you create it, you will always be working with a spline object named Shape01.*

| Source Object | Spline Path Name |
| --- | --- |
| DumShkHiL1 | PathShkHiL1 |
| DumShkHiL2 | PathShkHiL2 |
| DumShkHiL3 | PathShkHiL3 |
| DumShkHiL4 | PathShkHiL4 |
| DumShkHiR1 | PathShkHiR1 |
| DumShkHiR2 | PathShkHiR2 |
| DumShkHiR3 | PathShkHiR3 |
| DumShkHiR4 | PathShkHiR4 |
| DumShkLoL1 | PathShkLoL1 |
| DumShkLoL2 | PathShkLoL2 |
| DumShkLoL3 | PathShkLoL3 |
| DumShkLoL4 | PathShkLoL4 |
| DumShkLoR1 | PathShkLoR1 |
| DumShkLoR2 | PathShkLoR2 |
| DumShkLoR3 | PathShkLoR3 |
| DumShkLoR4 | PathShkLoR4 |

### Step 16. **Make dummies follow paths**

You need to make the top and bottom dummy object of each shock follow its correct spline path. Assign a Path Controller to the position track of each shock dummy object. Each dummy object needs to be told which path to follow to generate the correct position information.

Select the dummy object. Under the Motion panel, expand the Assign controller rollout.

Highlight Position:Path. Click on Pick Path.

Choose the correct path for the dummy object you're working on by pressing the <H> key and selecting the appropriate spline object from the list. Refer to the table below to make sure you're assigning the dummy objects to the correct paths.

| Path | Dummy Object |
|------|--------------|
| PathShkHiL1 | DumHiL1 |
| PathShkHiL2 | DumHiL2 |
| PathShkHiL3 | DumHiL3 |
| PathShkHiL4 | DumHiL4 |
| PathShkHiR1 | DumHiR1 |
| PathShkHiR2 | DumHiR2 |
| PathShkHiR3 | DumHiR3 |
| PathShkHiR4 | DumHiR4 |
| PathShkLoL1 | DumLoL1 |
| PathShkLoL2 | DumLoL2 |
| PathShkLoL3 | DumLoL3 |
| PathShkLoL4 | DumLoL4 |
| PathShkLoR1 | DumLoR1 |
| PathShkLoR2 | DumLoR2 |
| PathShkLoR3 | DumLoR3 |
| PathShkLoR4 | DumLoR4 |

## Step 17. **Display final model**

The animation is now ready to be rendered.

Unhide all the hidden objects.

Some objects aren't required for the rendering. Select all objects beginning with Dum, and hide them.

Save your work as VEHC08.MAX.

Create a camera, set up lights and render the animation. The vehicle bumps along, following every curve of the terrain.

To see animation created with these tutorials, see VEHHIREZ.AVI or VEHLOREZ.AVI in the \KYLE\AVI directory on the CDROM.

### Things to Try

The principles used here can be applied to virtually any vehicle with any number of wheels on almost any terrain.

There are many elements you could add to this animation. You might calculate the acceleration of the vehicle and write an expression, relating it to the burst rate of a particle system for sand or dust coming from under the wheels, or for smoke coming from an exhaust pipe. You can find an example of varying parameters based on another object's velocity in the file EXPVELOC.MAX located in your \3DSMAX\SCENES directory. This file comes with 3D Studio MAX.

The animation model you have created here will allow you to speed up and slow down at any time along the path. You can do this by adding an ease curve to the path percent track of DumBody. When you assign each wheel dummy object to its path, you can assign an Expression Controller to it's Path Percent value that looks at the DumBody Path Percent Controller. You can find an example of a wheel moving at varied speeds in the file EXPROLL4.MAX on the CDROM.

# appendix

## Resources

There are lots of places to go for more information on 3D Studio MAX and related products.

### Kinetix

Kinetix regularly releases new versions of products and provides many services such as training. Stay up to date by visiting their web page at

**www.ktx.com**

For information on upgrades, you can contact your 3D Studio MAX dealer. You can also call Kinetix directly at (800) 879-4233 or (415) 507-5000.

## Magazines

3D Artist
Columbine, Inc.
P.O. Box 4787
Sante Fe, NM 87502-4787
(505) 982-3532
(505) 820-6929 fax
info@3dartist.com
www.3dartist.com

CADalyst
Advanstar Communications, Inc.
P.O. Box 6136
Duluth, MN 55806-6136
(800) 346-0085 ext 477 or
(218) 723-9477
www.cadonline.com

NewMedia
HyperMedia Communications, Inc. (HCI)
901 Mariner's Island Road
Suite 365
San Mateo, CA
(415) 573-5170
(415) 573-5131 fax
edit@newmedia.com

DV (Digital Video)
ActiveMedia, Inc. and IDG Company
600 Townsend Street
Suite 170
East San Francisco, CA 94103
(800) 441-4403
(415) 522-2400
(415) 522-2409 fax
 www.dv.com

AV Video
Montage Publishing, Inc.
701 Westchester Avenue
White Plains, NY 10604
(800) 800-5474
(914) 328-9157
(914) 328-9093 fax
AVVideo@aol.com
www.kipinet.com

Computer Graphics World
PennWell Publishing Company
10 Tara Boulevard
5th Floor
Nashua, NH 03062-2801
(603) 891-0123
(603) 891-0539 fax
www.cgw.com

3D Design
Miller Freeman Company
600 Harrison Street
San Francisco, CA 94107
(415) 905-2200
www.3d-design.com

# 3D Studio MAX Buttons Listings

These button listings are provided for your convenience. Feel free to photocopy the following pages. A copy of these listings is also provided on the CDROM, in the file BUTTONS.DOC under the directory \MICHELE\BUTTONS. This file is in Word for Windows 6.0 format.

# Buttons in Order of Appearance

## Toolbar

| | | |
|---|---|---|
| Help | | |
| Undo | | |
| Redo | | |
| Select and Link | | |
| Unlink Selection | | |
| Bind to Space Warp | | |
| Select | | |
| Rectangular Selection Region | *flyout* | |
| Circular Selection Region | *from flyout* | |
| Fence Selection Region | *from flyout* | |
| Select by Name | | |
| Select and Move | | |
| Select and Rotate | | |
| Select and Uniform Scale | *flyout* | |
| Select and Non-uniform Scale | *from flyout* | |
| Select and Squash | *from flyout* | |
| Use Pivot Point Center | *flyout* | |

**Use Selection Center** *from flyout*

**Use Transform Coordinate Center** *from flyout*

**Restrict to X**

**Restrict to Y**

**Restrict to Z**

**Restrict to XY Plane** *flyout*

**Restrict to YZ Plane** *from flyout*

**Restrict to ZX Plane** *from flyout*

**Inverse Kinematics on/off toggle**

**Mirror Selected Objects**

**Array** *flyout*

**Snapshot** *from flyout*

**Align** *flyout*

**Normal Align** *from flyout*

**Place Highlight** *from flyout*

**Track View**

**Material Editor**

**Quick Render**

**Render Scene**

**Render Last**

## Panels

 **Create**

 **Modify**

 **Hierarchy**

 **Motion**

 **Display**

 **Utilities**

## Create panel

 **Geometry**

 **Shapes**

 **Lights**

 **Cameras**

 **Helpers**

 **Space Warps**

 **Systems**

## Prompt Line

🔒 **Lock Selection**

▦ **Crossing Selection**

▦ **Window Selection**

⬡ **Degradation Override**

▦ **Relative Snap**

$^2$ **2D Snap Toggle**                    *flyout*

$^{2.5}$ **2.5D Snap Toggle**              *from flyout*

$^3$ **3D Snap Toggle**                    *from flyout*

**Angle Snap Toggle**

$^%$ **Percent Snap**

**Spinner Snap Toggle**

## Time Controls

Key Mode Toggle

Go to Start

Previous Frame

Play Animation                              *flyout*

Play Selected                               *from flyout*

Next Frame

Go to End

Time Configuration

## Viewport Navigation

Zoom

Zoom All

Zoom Extents                          *flyout*

Zoom Extents Selected                 *from flyout*

Zoom Extents All                      *flyout*

Zoom Extents All Selected             *from flyout*

Region Zoom                           *ortho views only*

Field-of-view                         *non-ortho views only*

Pan

Arc Rotate                            *flyout*

Arc Rotate Selected                   *from flyout*

Min/Max Toggle

## Toolbar Buttons
## In Alphabetical Order

| | | |
|---|---|---|
| | **Align** | *flyout* |
| | **Array** | *flyout* |
| | **Bind to Space Warp** | |
| | **Circular Selection Region** | *from flyout* |
| | **Fence Selection Region** | *from flyout* |
| | **Help** | |
| | **Inverse Kinematics on/off toggle** | |
| | **Material Editor** | |
| | **Mirror Selected Objects** | |
| | **Normal Align** | *from flyout* |
| | **Place Highlight** | *from flyout* |
| | **Quick Render** | |
| | **Rectangular Selection Region** | *flyout* |
| | **Redo** | |
| | **Render Last** | |
| | **Render Scene** | |
| | **Restrict to X** | |

| | | |
|---|---|---|
| **Y** | **Restrict to Y** | |
| **Z** | **Restrict to Z** | |
| **XY** | **Restrict to XY Plane** | *flyout* |
| **YZ** | **Restrict to YZ Plane** | *from flyout* |
| **ZX** | **Restrict to ZX Plane** | *from flyout* |
| | **Select** | |
| | **Select by Name** | |
| | **Select and Link** | |
| | **Select and Move** | |
| | **Select and Non-uniform Scale** | *from flyout* |
| | **Select and Rotate** | |
| | **Select and Uniform Scale** | *flyout* |
| | **Select and Squash** | *from flyout* |
| | **Snapshot** | *from flyout* |
| | **Track View** | |
| | **Undo** | |
| | **Unlink Selection** | |
| | **Use Pivot Point Center** | *flyout* |
| | **Use Selection Center** | *from flyout* |
| | **Use Transform Coordinate Center** | *from flyout* |

## About the CDROM

The CDROM contains meshes, maps, AVIs, plug-ins and other files for use with this book.

When a file is called for during a tutorial, you are given the name of the file and the directory where it can be found.

No installation is necessary to use the files on the CDROM. When you want to use a file, you can copy it to your hard disk or simply load it directly from the CDROM.

The directories on the CDROM are named for each contributor. For example, you will find files for Kyle McKisic's tutorials under the \KYLE subdirectory tree.

Under the \EXTRA directory are meshes and AVIs from other master 3D Studio MAX artists not included in this book.

A full listing of the files on the CDROM can be found in the file README.TXT in the root directory of the CDROM.

# index